Dirty Secrets of Nuclear Power
in an Era of Climate Change

Doug Brugge • Aaron Datesman

Dirty Secrets of Nuclear Power in an Era of Climate Change

Foreword by Helen Caldicott

 Springer

Doug Brugge
University of Connecticut
Farmington, CT, USA

Aaron Datesman
University of Virginia
Washington, DC, USA

ISBN 978-3-031-59594-3   ISBN 978-3-031-59595-0   (eBook)
https://doi.org/10.1007/978-3-031-59595-0

This Springer imprint is published by the registered company Springer Nature Switzerland AG
The registered company address is: Gewerbestrasse 11, 6330 Cham, Switzerland

If disposing of this product, please recycle the paper.

Dedicated to the memory of Prof. Steven Wing of the University of North Carolina, Chapel Hill.

Foreword

This book, *Dirty Secrets of Nuclear Power in an Era of Climate Change*, is an imperative scholarly explanation of why nuclear power is NOT the answer to global warming.

It begins by carefully explaining the intricate mechanisms of global warming and the ongoing threat to biological life, and why some people think that nuclear power will be the obvious solution to this man-made dilemma.

The book then delves into the medical history and human tragedies associated with the entire nuclear fuel cycle beginning with uranium mining, to routine radioactive emissions from nuclear reactors to the tragedies of meltdowns, and on to the associated threat of nuclear war imposed by nuclear power, because countries can manufacture nuclear weapons from their nuclear power waste.

It then describes the tragedy imposed upon future generations from millions of tons of carcinogenic radioactive waste, which will pollute water and food chains for the rest of time, thereby inducing random compulsory genetic mutations.

And finally it contains a lucid and scientific explanation of so-called "low dose" radiation and its clear relationship to cancer, and the devious ignorance of this subject perpetrated on the general public by the bastions of the all-powerful nuclear industry.

President Emeritus of Physicians for Social   Helen Caldicott
Responsibility—1985 Nobel Peace Prize
Melbourne, Australia

Prologue

One of us (Brugge) interviewed James Hansen, one of the earliest people to warn about the dangers of climate change, in 1988. He was, at that time, director of the NASA Goddard Institute for Space Studies in New York City. Hansen had sprung on the national scene that year after his Congressional testimony about climate change. In many ways, his testimony was the opening shot in what has become decades of fear, concern, debate and controversy.

Fast forward to the present and Hansen remains a dogged advocate who continues to promote robust responses to climate change. Over the course of his career the harms have become more apparent, with increasing impact expected in the future. This year, 2023, he published a controversial paper, covered by the *New York Times*, claiming the speed of climate change is faster and its consequences more imminent than the scientific consensus (Wallace-Wells, 2023).

Despite our admiration of Hansen and being open to his concerns that climate change may be more severe than commonly recognized, we find ourselves in disagreement with him on one crucial aspect. He continues, in his most recent article, to see an important role for nuclear power in limiting climate change (Hansen et al., 2023). Unlike Bill McKibben, another stalwart in the discussions and struggles about anthropogenic driving of global temperatures, Hansen sees nuclear as a critical element in the fight.

We hope, through the material we present in this book, to convince you, the reader, that despite the substantial consequences climate change poses and the superficial appeal of nuclear power as part of the response, nuclear holds too many risks, is too expensive and too slow to come online to play a major role. We are not reflexively anti-nuclear. Instead we think the record of nuclear power leads, objectively, to the conclusion that it is not a desirable nor viable option.

The point of this book is threefold. First to engage and convince people who are unsure about the issue. Second, to provide anti-nuclear advocates with well-reasoned arguments in support of their position. But also, third, to argue with reasonable people, like Hansen, who are rational and science-based, that nuclear has limitations that they either ignore or underestimate.

We want to emphasize that we respect those who disagree with us based on their alternative interpretation of the evidence. However, we seek to present here only a non-technical critique of nuclear power so that it is accessible to non-scientists. Instead of amassing as much evidence in favor of our argument, we intend this book to acknowledge the limitations as well as the strengths of our position. We also understand that complete objectivity is impossible. However, we believe it is possible to be transparent and strive to do so.

If this book generates debate and discussion that convinces you to think more deeply and critically about the issue, it will have served its primary purpose.

Finally, before turning the reader over to the main text, it is worth noting that while we agree almost entirely in our critique of nuclear power relative to climate change, each of us has a distinct "voice" and approach to our writing. Since we took the lead on different chapters, the reader will notice our respective styles. To make the transition between chapters clearer, we have listed which of us was the lead author on each chapter.

References

Hansen, J. E., Sato, M., Simons, L., Nazarenko, L. S., Sangha, I., Kharecha, P., Zachos, J. C., von Schuckmann, K., Loeb, N. G., Osman, M. B., Jin, Q., Tselioudis, G., Jeong, E., Lacis, A., Ruedy, R., Russell, G., Cao, J., & Li, J. (2023). Global warming in the pipeline. *Oxford Open Climate Change, 3*(1). https://doi.org/10.1093/oxfclm/kgad008

Wallace-Wells, D. (2023). *The Godfather of climate science turns up the heat: David Wallace-Wells.* https://www.proquest.com/blogs-podcasts-websites/godfather-climate-science-turns-up-heat/docview/2887030899/se-2?accountid=30699

Acknowledgements

We thank Sangita Kunwar for her assistance with manuscript preparation, especially seeking permissions to reprint figures, and Sandy Bartholet for reviewing the material contained in Chapter 8. We also wish to thank our individual spouses, Miho Matsuda and Andria Thomas, for their unwavering support and patience. The authors furthermore acknowledge with gratitude the involvement and superlative effort of their collaborators on the 3MILER RUN investigation: Heidi Hutner of Stony Brook University, Susan Bailey of Colorado State University, and Chris Tompkins and Erin Robinson of KromaTiD, Inc. Open access publication of the eBook was made possible by funds from the Health Net, Inc. Endowed Chair in Community Medicine.

Contents

Chapter 1
Climate Change: Melting Ice and Statistical Models

It is not the purpose of this book to litigate the issues around climate change itself. Rather, we accept that climate change is happening and that it is caused primarily by human activity, even if the rate and severity might be open to debate. Nevertheless, a brief review of the state of climate change and the science behind it is appropriate before we delve into the thorny issues surrounding the possible role of nuclear power in averting the worst outcomes.

As we write, there remains a robust and often frustrating public debate about the reality and nature of climate change and whether it is driven by the processes of modern industrial society. The discussion, if you can call it that, is lopsided because the evidence weighs so heavily toward confirming anthropogenic climate forcing. Public discourse still manages to be, at times, quite acrimonious.

This controversy is reminiscent of other environmental issues that rest largely on scientific evidence. The push back against science that establishes the harm of products from industry was pioneered long ago by the tobacco companies. They honed their approach first to resist evidence of harm from smoking, then to suppress or delay concerns about second-hand smoke, by focusing on the evidence being less than perfect (Brugge, 2018).

Today, it is the fossil fuel companies that have an incentive to generate what David Michaels called "The Triumph of Doubt" about the science of

---

The original version of the chapter has been revised. A correction to this chapter can be found at https://doi.org/10.1007/978-3-031-59595-0_9

Doug Brugge is the primary author of this chapter.

© The Author(s) 2024, Corrected Publication 2024
D. Brugge, A. Datesman, *Dirty Secrets of Nuclear Power in an Era of Climate Change*, https://doi.org/10.1007/978-3-031-59595-0_1

climate change (Michaels, 2020). Because of this, public understanding of climate change is fraught on multiple levels. A motivation for writing this book is to seek to convey science and evidence clearly, at a level that is accessible and without distortion, acknowledges the limitations of the evidence, but sets a reasonable bar (rather than impossible) for making decisions.

A significant problem with public controversies that revolve around scientific questions is that the science can be difficult for untrained people to grasp, it can be manipulated by political actors and then dramatized by the media to grab attention. A sober, thoughtful, and serious conversation can be challenging to engage amid the swirling maelstrom of angry posts on social media and poorly translated or understood science news.

The issues related to climate change can be broken down into three parts. First, is the climate warming? Second, if it is, is the warming caused primarily by human activities and, in particular, the burning of fossil fuels? Third, provided it is we humans who are the underlying cause, how rapid is the change and, based on that, how much time do we have to adjust to avoid serious consequences?

We cannot delve deeply enough here to have a nuanced discussion of the science of climate change. Rather, we seek to stake our position prior to exploring in much more detail the possible role of nuclear power for slowing climate change. If the reader is, at this point, in need of convincing that climate change is real, anthropogenic and poses consequences within decades, we suggest they seek out that literature and digest it prior to reading this book (PCC SAR SYR, 1995; Trenberth & Cheng, 2022).

We consider first the melting ice. While ice melt is not as scientifically rigorous as modeling, it has a couple of advantages. First, it is highly visible, which makes it more compelling than numbers on a page or even a very clear graph. Second, while there are complexities to the processes by which climate change melts ice, the melting itself is a legitimate measure of integrated changes in temperature of air and water. Also melting ice is a more stable indicator than the weather which varies so much day to day and season to season. (Sengupta, 2023).

A key figure in documenting the melting ice is the underappreciated work of Bruce Molnia. After 42 years of service to the US Geological Survey, Dr. Molnia retired in 2019 from his position as Senior Science Advisor for National Civil Applications in the National Civil Applications Center. The core of his research career was studying the glaciers of Alaska. The title of his 2007 solo authored paper, "Late nineteenth to early twenty-first century behavior of Alaskan glaciers as indicators of changing regional climate",

Fig. 1.1 Two pairs of photographs showing how the glaciers changed over time. (**a**) Toboggan Glacier, June 29, 1909; and (**b**) on September 4, 2000. Both were taken from the same location in Harriman Fiord, Prince William Sound. (**c**) White Thunder Ridge, Muir Inlet, Glacier Bay National Park and Preserve, August 13, 1941, by William O. Field, and (**d**) August 31, 2004, by Sidney Paige. There is no vegetation in the 1941 photograph. The photographs document the significant retreat of the glaciers over many decades. (Reproduced with permission from Molnia, 2007)

explains why his research focus helped give climate change physical manifestations.

Changes, specifically "retreat", what we might commonly think of as melting, of glaciers was one of the earliest tangible signs of climate change. In 1999 US Secretary of the Interior Bruce Babbit asked Molnia to find "unequivocal" evidence of climate change. Molnia's paired photographs of glaciers in the past and present were his answer (Fig. 1.1; Molnia, 2007).

The science underlying glacier melting is not simple because there are other factors besides climate warming at play. However, in most cases it appears that glacier retreat is, indeed, secondary to climate change. There are a few cases of glaciers that are expanding, but that is rare and explained by the peculiar circumstances of those glaciers. In fact, it is possible to think of the glaciers as the canary in the coal mine, because they, like the birds that miners took with them into mine shafts, are early indicators of the physical effects of rising temperatures.

Unlike the polar ice sheets, which are massive, the glaciers are comparatively small and often adjacent to warmer regions of the globe. Thus, their melting is more readily apparent. The melting of glaciers is readily visible, an advantage over statistical models that are complex and not so easily rendered in easily understood images.

Despite being less visible and in some ways less dramatic, in the same timeframe that Molina was documenting the conversion of glacial termini into lakes, warming trends had also begun to eat away at the most vulnerable edges of sea ice in Antarctica. Between 1995 and 2002, large sections of the Larsen B Ice Shelf collapsed (Fig. 1.2; NASA Observatory, 2002) Not long after, the Wilkens Ice Shelf also began to deteriorate. Both are on the Antarctic peninsula, the most exposed and vulnerable ice on the continent.

There is yet another massive storage of ice that is more difficult to see than the glaciers and polar ice. This is the permafrost, essentially, frozen ground. Permafrost is found mostly in the arctic, but also at high elevations, notably the Himalaya Mountains in South Asia which is sometimes called the "Third Pole" because of its smaller, but still considerable, ice content. As permafrost melts it has revealed ancient remains of animals and plants that have been preserved in a frozen state for millennia (Fig. 1.3; Plotnikov, 2020).

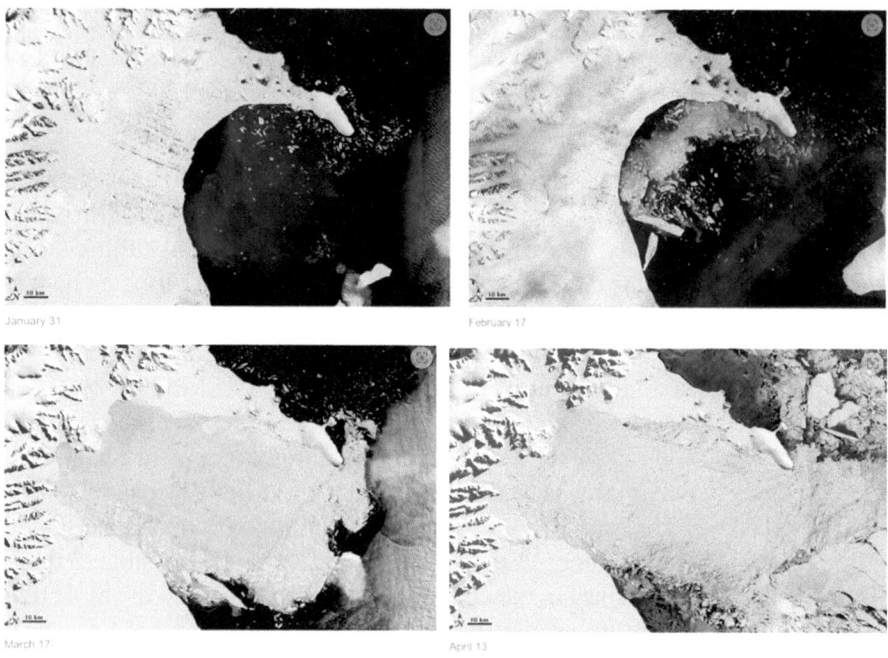

Fig. 1.2 Collapse of the Larsen B Ice Shelf on the Antarctic Peninsula from January to April 2002. The shelf is sea ice so more vulnerable to warming than ice on land. Also, because it is sea ice, it does not add to sea level rise. (Reproduced from NASA Observatory, 2002)

Fig. 1.3 A photograph of a largely preserved carcass of a woolly rhino that emerged from the melting permafrost in August 2020 in Yakutia, Russia. (Reprinted with permission from the Associated Press (Plotnikov, 2020))

Temperatures are rising faster in the arctic than at the Third Pole largely because the extensive cover of white ice reflects sunlight back before it is absorbed and warms the surface. Counter intuitively, ice is melting faster at the Third Pole, in part because the Arctic has boreal forests and moss coverage that the Third Pole does not.

As the ground ice melts it releases methane and carbon dioxide. An estimate of the amount of organic carbon in the soil in the Northern Hemisphere is 1700 Pg, about equal to the mass of all of the water in Lake Ontario. The melting of the permafrost creates a positive feedback loop in which more melting releases more carbon into the atmosphere, driving further increase in temperatures and then more warming of ground ice and more release of carbon. (Nisbet et al., 2023).

Sometime in the fall of 2013 a massive cylindrical crater formed in the Siberian permafrost. Scientists flew out from Moscow to examine this new feature in the earth and observed that it, and others found subsequently, were formed suddenly by violent explosions that thrust soil and ice hundreds of meters. There were signs of burning at the remaining edges of the craters. (Gray, 2020).

It is now established that these craters are created by blasts of methane gas. It appears that a warmer climate is releasing trapped methane in the frozen ground that builds up and forms a mound. After the pressure in the mound becomes too great, it is released in a blast that leaves a cylindrical crater, almost as if a round cookie cutter had excised a piece of the earth (Fig. 1.4; Pushkarev, 2014).

Fig. 1.4 This picture is of a crater in north-west Siberia on the Yamal Peninsula that is 164 feet-deep (50 m). The hole formed in 2013 and apparently was created by the explosion of methane gas. (Used with permission Reuters (Alaska Public Media, 2022))

If the melting ice provides a tangible indication of the impact of climate change so far on our planet, it cannot tell us what will happen in the future. For that we need modeling. By its nature, modeling is a highly technical exercise that in its full details is virtually impenetrable for the non-scientist. All modeling shares these features, but climate modeling, because of the consequences and the inherent complexity, is even harder to explain and assess.

Perhaps a comparison to the models with which we are most familiar is helpful. We all depend on these models because they predict the weather. Weather models, as we all know, have improved over time (they were too often wrong 40 years ago) yet still retain a degree of error. Usually they are reasonably accurate, but they have limits. If one watches the prediction a week ahead and pays attention as the day approaches, the prediction changes and, usually, becomes more accurate.

Weather models are both helpful and problematic when trying to explain climate change modeling. From an experiential standpoint, they can give the reader a general sense of what models are and how they function. However, unlike melting ice, weather is a poor metric by which to observe climate change. This is because weather is, in many places, highly variable. Weather can seem, erroneously, to confirm climate change during a heat wave and challenge it during a blizzard.

Climate change models use many variables as inputs—such as temperature, estimates of carbon dioxide releases, cloud cover, geography and others—to predict changes in climate variables, much as weather models predict temperature, precipitation and wind. Climate models are usually compared to data from the past to test their accuracy. There are many climate models, each with slightly different approaches and inputs, developed by teams of researchers that result in a range of outcomes and magnitudes of error.

Too often, popular debates about climate change revolve around whether it is real, a black and white absolutism that fails to reflect the underlying science. Instead, we would urge the reader to consider that the main debate is about how fast climate change is happening because that is what affects the scale and timing of responses that are needed.

We see that climate change is an impending crisis, but compared to, for example, the Fukushima nuclear accident, it is a slow-moving disaster, unfolding over decades. Because climate change is a gradual accumulation of gasses in the atmosphere that contribute to warming and because these gasses have long lives, reversing climate change will also be slow. There is already substantial momentum forcing temperature rises that will not be possible to reverse quickly.

While tracking the changes in global temperature and observing their more obvious impacts, including changes in the ice, has some challenges, it is comparatively straightforward relative to predicting the future of climate change.

In the context of climate change, the input data and temporal and spatial scale of modeling is much larger than models that attempt to predict the weather a day or a week from now. Climate models can be global in scope and seek to predict what will happen decades from now. Thus, climate models require vast amounts of computer capacity to make calculations based on immense computer codes. Inputs into these models include solar radiation, concentrations of gasses in the atmosphere that increase temperature (such as carbon dioxide and methane) and concentrations of particulate matter in the air that reduce temperature.

No models are perfect. Therein lies the real potential controversy about climate change. Different models offer different predictions of the trajectory of climate change. Some predict that we have more time, others that we have less. If the models predicting slower change are more accurate, we have more time to adapt. However, since we cannot be certain, we think that it would be a mistake to assume the best-case scenario is correct (Fig. 1.5). If we are wrong in our optimistic assumption, we will have even less time to respond once we figure that out and the costs will be greater and the damage more severe.

It is better to prepare for the worst and if that is overly cautious, there will be many ancillary benefits to burning less fossil fuel. The primary of which will be reducing ambient air pollution. Particulate matter air pollution, which derives substantially from combustion sources, is responsible for millions of deaths worldwide every year and even more serious chronic illnesses (Chang et al., 2022). Frankly, the toll from air pollution that derives

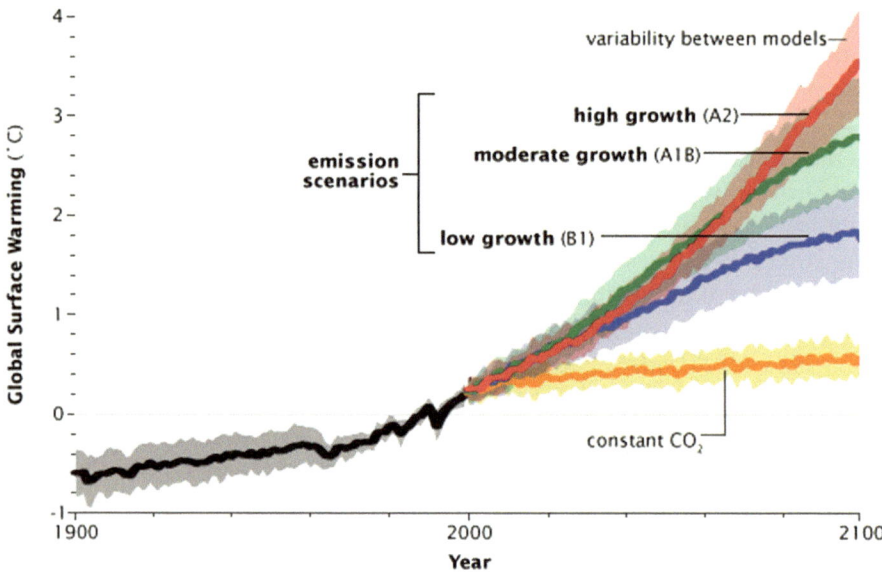

Fig. 1.5 NASA prediction of global temperature increases depending on changes in CO_2 emissions. (NASA Earth Observatory, n.d.)

from the same sources as climate change should, by itself, justify moving away from combustion related climate forcing emissions.

For our purpose in this book, time is a significant factor in terms of considering nuclear power as a response to climate change. The problem for nuclear power supporters is that in the current context, at least in highly developed countries, approving, financing and building nuclear power plants takes an inordinately long time. We will discuss this in detail in Chap. 8. Since we need to respond to the climate change threat quickly, a power source that is slow to come online is unlikely to be a viable part of our response.

Summary Points

1. Melting glaciers and polar ice are highly visible and reasonably accurate indicators of climate change.
2. Statistical models of global temperature change show consistent increases, although the speed with which warming is happening varies based on the model assumptions.
3. The need to move away from fossil fuels raises the question of which sources of energy are best, including the possible role of nuclear power.

References

Alaska Public Media. (2022). *Alaska permafrost thaw is a clue in the mystery of Arctic methane explosions.* https://alaskapublic.org/2022/02/03/alaska-permafrost-thaw-is-clue-in-mystery-of-arctic-methane-explosions/

Brugge, D. (2018). *Particles in the air: The deadliest pollutant is one you breathe every day.* Springer. https://doi.org/10.1007/978-3-319-89587-1

Chang, A. Y., Chatterji, S., Abedi, A., Afarideh, M., Ahmadi, M., Almadi, M. A. H., Amit, A. M. L., Andrei, T., Asadi-Pooya, A., Balachandran, A., Banach, M., Bedi, N., Bell, M. L., Bhattarai, S., Bhutta, Z. A., Bibi, S., Bolla, S. R., Briggs, A. M., Cámera, L. A., et al. (2022). Global, regional, and national burden of diseases and injuries for adults 70 years and older: Systematic analysis for the global burden of disease 2019 study. *Bmj, 376*, e068208. https://doi.org/10.1136/bmj-2021-068208

Gray, R. (2020). The mystery of Siberia's exploding craters. *BBC.com* https://www.bbc.com/future/article/20201130-climate-change-the-mystery-of-siberias-explosive-craters

Michaels, D. (2020). *The triumph of doubt: Dark money and the science of deception.* Oxford University Press.

Molnia, B. F. (2007). Late nineteenth to early twenty-first century behavior of Alaskan glaciers as indicators of changing regional climate. *Global and Planetary Change, 56*(1–2), 23–56. https://doi.org/10.1016/j.gloplacha.2006.07.011

NASA Earth Observatory. (n.d.). *Global warming.* NASA Earth Observatory. Retrieved October 10, 2023, from https://earthobservatory.nasa.gov/features/GlobalWarming/page5.php

NASA Observatory. (2002). *World of change: Collapse of the Larsen-B ice shelf.* NASA Observatory. https://earthobservatory.nasa.gov/world-of-change/LarsenB

Nisbet, E. G., Manning, M. R., Dlugokencky, E. J., Michel, S. E., Lan, X., Röckmann, T., van der Gon, D., Hugo, A. C., Schmitt, J., Palmer, P. I., Dyonisius, M. N., Oh, Y., Fisher, R. E., Lowry, D., France, J. L., White, J. W. C., Brailsford, G., & Bromley, T. (2023). Atmospheric methane: Comparison between Methane's record in 2006–2022 and during glacial terminations. *Global Biogeochemical Cycles, 37*(8), n/a. https://doi.org/10.1029/2023GB007875

PCC SAR SYR. (1995). *Climate change 1995; impacts, adaptations, and mitigation.* Contribution of working group II to the second assessment report of the Intergovernmental Panel on Climate Change; summary for policymakers.

Plotnikov, V. (2020). *Woolly rhino? Ice age carcass recovered from permafrost in Siberia.* The Associated Press. https://www.oregonlive.com/environment/2020/12/woolly-rhino-ice-age-carcass-recovered-from-permafrost-in-siberia.html

Pushkarev, V. (2014). *Siberia's massive permafrost craters may be result accumulation of methane gas due to climate change.* Reuters. https://www.news18.com/news/buzz/siberias-massive-permafrost-craters-may-be-result-accumulation-of-methane-gas-due-to-climate-change-3446924.html

Sengupta, S. (2023, September 13,). Climate change is melting Mount Rainier's glaciers. *The New York Times* https://www.nytimes.com/2023/09/12/climate/mount-rainier-glaciers-climate-change.html?smid=nytcore-ios-share&referringSource=articleShare

Trenberth, K. E., & Cheng, L. (2022). A perspective on climate change from Earth's energy imbalance. *Environmental Research: Climate, 1*(1), 13001. https://doi.org/10.1088/2752-5295/ac6f74

Chapter 2
The Dirty, Working-Class Problem

When people think of nuclear power, they usually associate it with sophisticated nuclear reactors and their adjacent cooling towers. Despite being over 70 years old, nuclear technology remains a prime example of modern, high-tech engineering and a sign of the advanced state of our civilization. But the uranium fuel rods that are at the center of the controlled nuclear reaction that boils water for steam, begin as ore, dirt and rock. Buried in the ground, uranium must be mined and processed before becoming the fuel that can sustain a nuclear reaction.

The mining and milling of uranium ore into yellow cake and conversion of yellow cake into a form that can be enriched for use as fuel, was, and still is in most of the world, a dirty and dangerous process for the workers. It is also one that contaminates adjacent land and water putting the families of the workers and local communities at risk. The scale of this problem is small compared to climate change, but as will be apparent in this chapter, the impact to many local communities, especially indigenous communities, can be devastating.

As a clarifying example, we will focus on the United States here, then briefly expand to a global context after that. Uranium mining in the US picked up after 1948 when the US Government assured a price for the ore and designated itself as the sole purchaser. By the late 1950s, the mining of uranium was a growing industry (Fig. 2.1), concentrated in the Southwestern US on the Colorado Plateau, where it remained active into the 1980s.

In a pattern that was repeated in many other countries and recapitulated today in other parts of the world, mining encroached substantially on indigenous communities. Most prominently in the US, the Navajo People were drawn into mining. For many of them it was their first experience with wage

Doug Brugge is the primary author of this chapter.

© The Author(s) 2024
D. Brugge, A. Datesman, *Dirty Secrets of Nuclear Power in an Era of Climate Change*, https://doi.org/10.1007/978-3-031-59595-0_2

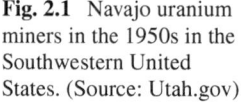

Fig. 2.1 Navajo uranium miners in the 1950s in the Southwestern United States. (Source: Utah.gov)

labor. Pay was low, but for people who had not had previous access to money, access to cash was appealing. (Brugge & Goble, 2002).

George Tutt, a Navajo man who had worked in the uranium mines described the rudimentary methods and conditions that constituted the first step of the process that led to nuclear bombs and power, "… we built tunnels and hauled it [uranium] out. We used wheelbarrows, shovels, and picks. They were the only tools we used."

Nor were they provided information about the health risks. He goes on to say, "We thought we were very fortunate, but we were not told, 'later on this will affect you in this way.' True, the men worked. When work stopped at the end of the shift, they just got out of the mines and went straight home. They were not told to wash or anything like that." (Miller, 2007). Of course, the most significant hazard in underground mines was radon gas and its radioactive offspring, which were invisible and odorless, followed by silica dust, which was readily apparent.

As in other countries that engaged in mining uranium in the post-World War period, the US, led by our Public Health Service, conducted a longitudinal epidemiology study of the miners. The goal of the study was not to determine whether the radon would cause lung cancer, as that had been firmly established by the late 1940s, but rather to estimate the dose response relationship. Miners enrolled in the study were not told of the risks they were unknowingly taking (Brugge & Goble, 2002).

The entire mining operation was cloaked in national security concerns that superseded consideration of the health of the miners or damage to the environment with its resulting risk to their families and communities. By 1959 the PHS study had shown a statistically significant association between radon exposure and lung cancer, as was expected, but a federal standard limiting radon exposure did not go into effect until 1969, until far too late to protect thousands of miners.

As miners became aware of the harms they were suffering, most obviously lung cancers at an early age, including among non-smokers, they began to ask questions and eventually, overcoming a steep learning curve, organized to seek compensation. The Navajo people took the lead in the campaign, trying first to sue the US Government. When that was blocked in court, they sought redress in Congress.

The initial injustice of the mining was compounded by what became a two-decade long campaign before the US Congress passed the Radiation Exposure Compensation Act in 1990. RECA extended monetary compensation to former miners and their families. Sadly, RECA contained within it further unfairness. The qualifying criteria in 1990 RECA were so stringent that they excluded many deserving miners and their families. In essence, the law, and with it the US Government, rubbed salt in still open wounds of many former miners and their families.

Another decade of advocacy was necessary to correct most, but even then, not all of the shortcomings of the original RECA legislation. The new RECA law, which went into effect in 2000, lowered the threshold for eligibility to a doubling of lung cancer risk and expanded eligible workers to include mill workers and above ground miners. It also moved oversight from the Department of Justice, which was perceived to be indifferent to workers, to the Department of Labor.

The accounting of compensation that RECA eventually paid out provides a record of the toll of death and illness left behind by uranium mining in the US. According to the US Department of Labor, as of 2024, benefits have been provided to more than 9000 uranium workers for a total of over $900 million (Table 2.1). The number of workers compensated is an important value because it is a conservative estimate of the harm mining uranium caused in one country.

This estimate of health consequences is conservative because the criteria for compensation are narrow and strict. A successful applicant must have been diagnosed with one of the following diseases: primary lung cancer; fibrosis of the lung; pulmonary fibrosis; cor pulmonale related to fibrosis of the lung; silicosis; or pneumoconiosis (also renal diseases for mill workers). In addition, their work history must either be calculated to exceed a doubling of risk for lung cancer, or they must have worked for at least 1 year.

Table 2.1 Radiation exposure compensation: claims to date summary of claims Received by 1/30/2024

Claim type desc	Pending	Approved	% of approved/ of disposed	$ approved	Denied	Total
Downwinder	115	26,463	84.1	$1,323,120,00	5013	31,591
Onsite participant	34	5569	58.9	$407,111,952	3881	9484
Uranium miner	33	6961	62.6	$695,374,560	4165	11,159
Uranium Miller	7	1945	74.4	$194,500,000	671	2623
Ore transporter	4	416	71.0	$41,600,000	170	590
Total	193	41,354	74.8	$2,661,706,512	13,900	55,447

Source: Civil Division (2024). Awards to date as of 01/31/2024. www.justice.gov. Retrieved February 2, 2024, from https://www.justice.gov/civil/awards-date-01312024

These criteria likely underestimate the toll for two primary reasons. First, lung cancer is probabilistic which means that many workers developed lung cancer from uranium mining at exposures below the doubling of risk threshold. That is, at half the doubling of risk exposure, one-third of the lung cancers would be expected to be caused by radon in the mines. Workers at that level of exposure are not eligible for compensation. It is even possible that most of the lung cancer was caused by exposures below the cut-off because so many workers had low exposures. Second, the assignment of exposure has considerable error as it is based on air monitoring in a tiny fraction of mines.

We humans are not, however, moved as much by statistics as we are by personal stories. Minnie Tsosie, a Navajo woman who was left a widow when her husband, who had worked in the mines, relates her husband's illness:

> Some years after [working another job since mining] he suddenly started having fevers quite frequently. At night he would get feverish, he said he thought his bones would ache…. it continued like that for many years. There was a time when it was like that and I never paid too much attention, until one time I started telling him to have a doctor check him….Then he went to see a doctor and he was told that the pain that he was feeling was caused from the mine work he had done. He was given pills and thereafter he took the pills. That made things better for a short while and then he would feel bad again. Then he got worse and it did not take long after that, not many years, it immediately brought his life down. From the time he was at his worst it was less than a year and he died. It did not take long. (Brugge & Goble, 2002)

The US example is bad enough by itself, but, sadly, it is only one of many worldwide. The National Academies of Sciences, in its 1996 report on the risks of radon exposure, cites cohort studies of lung cancer in uranium miners from seven counties: the US, France, China, Sweden, Czechoslovakia, Canada, and France (NRC, 1999). Uranium mining historically affected miners in many additional countries, notably in the former East Germany, for whom a record of the consequences is not as readily available.

Thus, the deaths and illnesses in US workers are a small fraction of the global health burden from mining uranium. It has sometimes been claimed that nuclear power either killed no one or very few people. The only way to make that claim credibly is to ignore the mining, milling, and processing of uranium and the tens of thousands of deaths and illnesses that resulted.

Today, mining in high income countries has mostly ended. The main exceptions being Australia and Canada. In the US and EU, mining of uranium is currently rare to non-existent, albeit with occasional, largely unsuccessful, efforts to revive it. However, the decline of active mining was not the end of the story because most of the inactive mines and mills became hazardous waste sites that required remediation.

Decades later, these sites continue to present a threat to the health of people living nearby or spending time on them. As with the impact of uranium mining on the health of miners themselves, the legacy of abandoned mines and mills is a global concern, with similar stories spread around the many countries that engaged in uranium mining. Disturbingly, the association with indigenous and tribal peoples is also replicated in many countries, including the First Nations in Canada and Aboriginal People in Australia.

While remediation of abandoned mines has progressed slowly in the US, remediation of mills was undertaken by the Uranium Mill Tailings Radiation Control Act of 1978 (UMTRCA), resulting in former mill sites being largely under control (Lohmann, 2022). "By August 1999, remedial actions were completed at 18 sites ... Those sites are now subject to long-term care and maintenance under the general NRC license (U.S.NRC, 2017). The cost of this to the US taxpayers was over $2 billion and will require ongoing site monitoring and maintenance essentially forever (Fig. 2.2).

Fig. 2.2 A uranium mill tailings remedial site. (Source: Doug Brugge)

For people who experienced the deaths of many, in some small communities a majority, of the men who worked in the mines, it was a completely reasonable concern that the families might also have been exposed and therefore at risk. George Lapahie, a Navajo man, and former miner, said in a 1995 interview:

> Today, we are experiencing a great amount of problems. That is what happened to my children. They have tumor problems. What is it coming from? Through their investigation they have traced it to the uranium. One had serious work done on his head. Their skull was cut and had radiation treatment. That is how it is. Another was affected in their internal organs. My sons and daughter are like that. Where is this coming from? In the past there were never stories about this. Now, those of us who have worked with uranium, our children are affected by it. In Shiprock, there was a big pile of it. [The children] used to go over there because we [lived] nearby. The houses which were on this side of it, I bought a home there. From there I went to work. They used to ride their bikes on the tailing pile to play and now it is like that today. (Brugge & Goble, 2002)

It is unlikely that radon is the causal agent of cancers for people exposed in the community, as opposed to the workplace, because radon disperses easily outdoors. Instead, the leading concern is solid radioisotopes, such as uranium and radium, and non-radioactive elements such as arsenic, all of which are found in high concentrations in uranium ore and could be ingested in drinking water or food or inhaled in dust.

Of course, Mr. Tutt's observations do not constitute scientific proof that mining was the cause of his children's illnesses. Rather, Mr. Tutt's suspicions contribute to a hypothesis that deserved, and still deserves, to be researched. Fortunately, over the decades since the interview with Mr. Tutt, research has slowly advanced in this area.

The University of New Mexico, in collaboration with community-based organizations, has spearheaded several major research efforts on the possible effects of environmental (non-occupational) exposure to uranium mine waste. One critical finding from their research was that among over 1300 Navajo people, those living in close proximity to uranium mine features had greater kidney disease, diabetes and hypertension (Hund et al., 2015). Proximity is a relatively crude measure of exposure, so additional research is needed, but their findings substantiate that there is, indeed, reason for concern.

A 2007 study led by a Navajo woman at Northern Arizona University, another center for this line of research, exposed mice to environmentally realistic concentrations of uranium in their drinking water. She found that the exposure led to, "reduction of primary follicles, increased uterine weight, greater uterine luminal epithelial cell height, accelerated vaginal opening, and persistent presence of cornified vaginal cells" all indicators of exposure to compounds that mimic the hormone estrogen. Further, adding a molecule that blocks estrogen activity, prevented the changes induced by uranium

(Brown, 2007). Animal studies are not human studies, but the findings were, and remain in our opinion, worrisome. There has been little follow-up on this result to our knowledge.

Monitoring of drinking water in the Navajo region has also repeatedly documented contamination by not only uranium, but also arsenic, which is often found in the same ore (Blake et al., 2015). Because many residents of the Navajo Nation use unregulated water sources originally intended only for livestock, there is increased risk of ingestion and elevated exposures.

While the level of health impact to community members is not as dramatic or easy to document as lung diseases in underground miners, the evidence so far is worrisome and suggests more should be done to reduce these exposures.

In the last decade, uranium mining in Central Asia and Africa, specifically in the countries of Kazakhstan and Namibia, but also Uzbekistan and Niger, has eclipsed output from Canada and Australia (World Population Review, 2023). The shift is hardly surprising. As the costs and consequences of uranium mining have become more apparent in high income countries, partly because regulations are more stringent, mining companies have sought friendlier locations that cost less and impose fewer restrictions.

One of us (Brugge) has traveled to Africa three times to participate in conferences there about uranium mining. In presentations, conversations, and site visits, it was obvious that mining on the African continent is largely replicating the laissez faire approach to mining in the US and other developed countries in the decades after World War II. Workers have little protection or knowledge of the risk. Control of environmental contamination is limited, exposing nearby communities. In South Africa, an informal settlement was visited that sat atop mine waste. It is clear to us that mine companies are taking advantage of low-income nations in Africa and that this trend is more likely to continue, and even expand, in the coming years (Winde et al., 2017; Fig. 2.3).

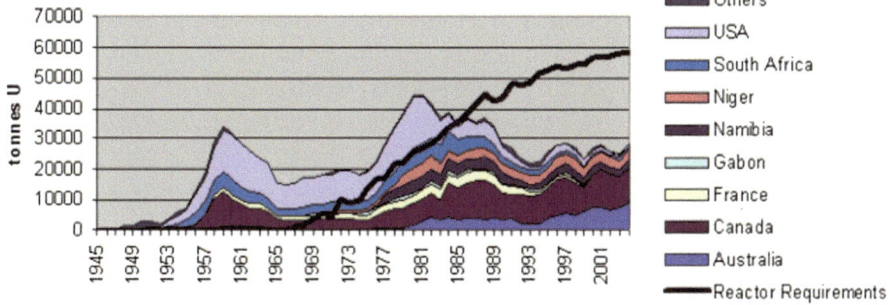

Fig. 2.3 Uranium production and use from 1945 to 2004. (Source: Winde et al., 2017)

Similarly, a town in Tajikistan that is adjacent to an abandoned uranium mine also revealed uncontrolled piles of mine debris and elevated radiation in soil and water (Dustov et al., 2013). There is every reason to believe, given the ease with which these sites were identified, that similar situations exist in many places in Central Asia and Africa and, in all likelihood, elsewhere in developing countries.

It is a sorry statement of the primacy of profits over concern for the health and well being of people, that so little is being done to address the ongoing tragedy of uranium mining. To us, there are two pressing needs. First, and most urgently, there is a need to extend first world occupational safety and environmental regulations to uranium mining in developing countries. This will raise the cost of mining uranium, but that cost is a tiny fraction of the cost of generating electricity with nuclear power.

Second, we need to finish research that is underway to more precisely establish the risk from environmental exposures, both from ongoing mining and legacy mining in countries such as the US where mining has largely concluded.

A longer-term goal is to force the externalized costs of uranium mining into the costs incurred by mining companies. Despite some settlements that assessed costs to the companies that were responsible, in the US we taxpayers were left to pick up too much of the cost of remediation of contaminated sites and long-term maintenance of stored uranium waste.

Summary Points

1. Often ignored when considering the health and environmental impacts of nuclear power are the early stages of the industry, including mining, milling and processing uranium ore.
2. The health and social impact of mining and processing uranium is considerable and has frequently had disproportionate effects on indigenous communities.
3. Although mining uranium is in decline in most high income countries, it continues to be conducted in poorly regulated and hazardous ways in low income countries, especially in Africa and Central Asia.

References

Blake, J. M., Avasarala, S., Artyushkova, K., Ali, A. S., Brearley, A. J., Shuey, C., Robinson, W. P., Nez, C., Bill, S., Lewis, J., Hirani, C., Pacheco, J. S. L., & Cerrato, J. M. (2015). Elevated concentrations of U and co-occurring metals in abandoned mine wastes in a northeastern Arizona native American community. *Environmental Science & Technology, 49*(14), 8506–8514. https://doi.org/10.1021/acs.est.5b01408

Brown, V. J. (2007). Uranium in drinking water: Low dose acts as endocrine mimic. *Environmental Health Perspectives, 115*(12), A595. https://doi.org/10.1289/ehp.115-a595a

Brugge, D., & Goble, R. (2002). The history of uranium mining and the Navajo people. *American Journal of Public Health, 92*(9), 1410–1419. https://doi.org/10.2105/AJPH.92.9.1410

Dustov, A., Mirojov, G., Yakubova, M., Umarov, S., Ishankulova, D., Eliasziw, M., & Brugge, D. (2013). Uranium mine proximity, immune function, and helicobacter pylori infection in Tajikistan. *Journal of Toxicology and Environmental Health, Part A, 76*(22), 1261–1268. https://doi.org/10.1080/15287394.2013.836694

Hund, L., Bedrick, E. J., Miller, C., Huerta, G., Nez, T., Ramone, S., Shuey, C., Cajero, M., & Lewis, J. (2015). A Bayesian framework for estimating disease risk due to exposure to uranium mine and mill waste on the Navajo nation. *Journal of the Royal Statistical Society. Series A (Statistics in Society), 178*(4), 1069–1091. http://www.jstor.org/stable/43965784

Lohmann, P. (2022). *Federal program—With no money attached—Created to tackle uranium mine cleanup.* Source New Mexico. https://sourcenm.com/2022/03/08/federal-program-with-no-money-attached-created-to-tackle-uranium-mine-cleanup/

Miller, A. C. (2007). *The Navajo people and uranium mining.* National Institute of Environmental Health Sciences. National Institutes of Health. Department of Health, Education and Welfare. https://doi.org/10.1289/ehp.115-a224a

NRC. (1999). *Health effects of exposure to radon: BEIR VI, committee on health risks of exposure to radon (BEIR VI).* (). National Academy Press.

U.S.NRC. (2017). Title, I program. *United States Nuclear Regulatory Commission.* U.S.NRC. https://www.nrc.gov/materials/uranium-recovery/regs-guides-comm/title-i-program.html

Winde, F., Brugge, D., Nidecker, A., & Ruegg, U. (2017). Uranium from Africa—An overview on past and current mining activities: Re-appraising associated risks and chances in a global context. *Journal of African Earth Sciences, 129*, 759–778. https://doi.org/10.1016/j.jafrearsci.2016.12.004

World Population Review. (2023). *Uranium production by country 2023.* World Population Review. Retrieved July 19, 2023, from https://worldpopulationreview.com/country-rankings/uranium-production-by-country

Chapter 3
Nuclear Waste

You only have a brief production of energy, but future generations are going to be grappling with waste forever.
– Gordon Edwards, President, Canadian Coalition for Nuclear Responsibility

Finland is a rare country that has embraced reliance on nuclear power and may be the first to complete a high level, long term nuclear waste disposal site (El-Showk, 2022). The repository, slated to open, if all goes as planned, in 2024 or 2025, is on the island of Olkiluoto, near Finland's west coast, facing the Gulf of Bothnia. Sedeer El-Showk, writing in *Science* (El-Showk, 2022), suggests that Finland's success at siting and building rests primarily on the socio-political context of the country. Finland is a country that rarely produces dissidents. Plus, there were considerable economic benefits offered to the host community. Fig. 3.1 shows a schematic of the design.

What is most striking from a global perspective, is how late and unusual is the possible success in Finland. The world's first commercial nuclear power plant began operation long ago—in December 1957—in Shippingport, Pennsylvania. Thus, we have been waiting more than 60 years for solutions to the disposal of high-level waste.

During that time, nuclear waste has been accumulating in dry cask storage at nuclear power plants around the US and the world. Let's make no mistake about how hazardous this waste is and how long it will take for it to decay to levels of radiation that are acceptable. The US Nuclear Regulatory Commission, far from the most alarmist about nuclear power risks, reports that, "10 years after removal from a reactor, the surface dose rate for a

---

Doug Brugge is the primary author of this chapter.

© The Author(s) 2024

D. Brugge, A. Datesman, *Dirty Secrets of Nuclear Power in an Era of Climate Change*, https://doi.org/10.1007/978-3-031-59595-0_3

Fig. 3.1 A schematic of the high level nuclear waste depository in Finland. GRAPHIC: V. ALTOUNIAN/SCIENCE

typical spent fuel assembly exceeds 10,000 rem/hour – far greater than the fatal whole-body dose for humans of about 500 rem received all at once." (El-Showk, 2022).

Ultimately there will have to be effective approaches to storing this waste for centuries. The waste is not going away and takes that long to decay to levels that are not immediately hazardous to health. It is important to recognize that science does not have a way to stop a radioactive substance from continuing to be radioactive. In other words, we cannot shut off radioactivity once we make radioactive elements in a reactor.

It might be helpful to explain why nuclear power produces this dangerous waste. The source of most of the radioactive byproducts originates from a subatomic particle called a neutron hitting an atom of uranium-235. U-235 is the rare isotope of uranium that is needed for nuclear fission. When U-235 absorbs a slow-moving neutron, it splits into two pieces. The pieces are called fission products. There are hundreds of different kinds of fission products produced within a nuclear reactor as the uranium atom splits in different ways. Fission also produces more neutrons which, in turn, split more uranium atoms leading to the escalating nuclear chain reaction.

In addition to heavy radioactive atoms, nuclear reactors also produce tritium. Tritium is a radioactive form of hydrogen with a half life of 12 years. Disposal of tritium is its own problem, since it is very difficult and expensive to separate tritium from the water in which it forms. Because of this, proposals have been floated to release it into the ocean or evaporate it into the atmosphere. A better approach would be to store it for 100 years in glass containers until its radioactivity is reduced by 99%.

Almost certainly the most important radioisotope in nuclear waste is plutonium. Plutonium is not a fission product as it is heavier than uranium and formed by a different process. Plutonium forms when an atom of U-238 (the more common isotope of uranium than U-235) absorbs a neutron, rather than splitting. When it does this, it turns into Pu-239. Plutonium is a critical byproduct because it can be used to make nuclear weapons. It was the ingredient in the bomb that destroyed Nagasaki and has been used extensively in nuclear weapons since then.

Gordon Edwards, quoted at the start of this chapter and whose work substantially informed this chapter, is President of the Canadian Coalition for Nuclear Responsibility (Gordon 2023). He has translated many of the issues surrounding nuclear waste into clear and understandable terms notably saying, "in exchange for, let's say, three generations of electricity, we have 300,000 generations of nuclear waste." From that perspective, he notes that nuclear waste is the main product of nuclear power and electricity is just a small blip early on (Gordon 2023).

The Nuclear Waste Management Association, which has a more optimistic view of nuclear power, characterizes the state of progress for disposal of high-level nuclear waste in some notable countries on its web site (NMWO, 2023). They note that Sweden appears close to having a site that can be developed for disposal of nuclear power waste. Indeed, in January 2022, Sweden announced that it had approved plans for a facility in Forsmark, 80 miles north of Stockholm. The Swedish plan is very similar to that of its neighbor, Finland. Approval may have benefited from similar levels of trust between the government and its population.

Other countries have struggled with gaining approval of host communities. France, a heavily nuclear country, has proposed using a site outside a village called Bure in the Champagne-Ardenne region in the eastern part of the country. If approved, and it is still faces political opposition, construction might begin in 2027, although that date is later than had previously been estimated (Mallet Benjamin, 2023).

In 2021, the United Kingdom, another country with substantial energy production from nuclear power, formed partnerships with two communities in Copeland, Cumbria that will involve discussions about possible disposal of highly radioactive, long-lived nuclear waste. This appears to be at an early stage, with the outcome not possible to predict yet.

Japan, a country that has also depended heavily on nuclear power prior to the Fukushima meltdown that led to closure of most of the country's nuclear reactors, also faces challenges with disposing of its nuclear waste. The risk of earthquakes and tsunamis are high on the Japanese islands. Selection of a site continues in Japan, with a desire to select one by 2025 and begin operation by 2035. This seems optimistic given experience in other countries to date.

The United States has perhaps the most dismal record of any nuclear country in terms of identifying a site for disposal of high-level nuclear waste. The US does have a low-level underground waste disposal site in southern New Mexico. But that site had a fire and emergency evacuation in 2014 (Gordon 2023). The cause of the fire was a chemical reaction of low level radioactive waste with kitty litter that resulted in a drum exploding and plutonium dust traveling more than 700 meters to the surface, contaminating 22 workers.

The process in the US for choosing and beginning construction on a repository for high level waste has cost billions of dollars over decades.

Despite the investment, the effort to find a viable option failed and none is likely for many more years.

Yucca Mountain in Nevada was the location of choice in the US. It was close to the nuclear test sites at which dozens of nuclear explosions had been detonated, first above ground, then below. But the goal of a repository at Yucca Mountain ran headlong into opposition by the State of Nevada and its powerful congressman, Harry Reid, Senate Majority Leader from 2007 to 2015. It may also have floundered by affecting the nearby lands of the Western Shoshone and Paiute Indians, once again (see Chap. 2) trying to impose nuclear risks on Native Americans.

Thus, today the process of identifying a site in the US and developing it is essentially at a standstill. It is rather amazing and disturbing that a country with 92 nuclear reactors harboring 88 metric tons of high-level waste has come full circle and is back to square one. The most recent siting process was canceled during the administration of President Trump. Apparently, the US Government is, as of 2023, "reviewing options and developing a new plan" (NMWO, 2023).

Reports about decay of high-level nuclear waste are often framed as time to reach a "safe" level of radiation. But "safe" is either an absolute elimination of risk, which is rarely, if ever, possible, or a relative metric based on one's values, essentially a low risk that we consider acceptable. In practice many assessments use the natural radiation of uranium ore as the benchmark for the level at which nuclear waste would no longer require stringent containment measures.

By this standard, it would take about 100,000 years for the waste to be comparable to natural uranium (Fig. 3.2), because natural uranium releases a low level of radiation, resulting from its long half-life (Corkhill & Hyatt, 2018). This time scale should be worrisome to the reader as it exceeds by an order of magnitude human civilization and by almost two orders of magnitude our modern technological progress with machines, industry, motor vehicles and rockets. We have often failed to predict dramatic problems with our technology, including impact on climate, that only became apparent in recent decades.

To provide context, consider that one hundred thousand years ago humans had recently migrated out of Africa and Neandertals still roamed what is now Europe. Travel was by foot and tools were simple and made of stone. Our ancestors lived in caves and other simple dwellings. And change was slow, very slow, compared to the sometimes-bewildering developments in technology we see today. Modest innovations took thousands of years.

Now, try to extrapolate forward in time and envision what our nuclear depositories might look like, how they would hold up, and whether they

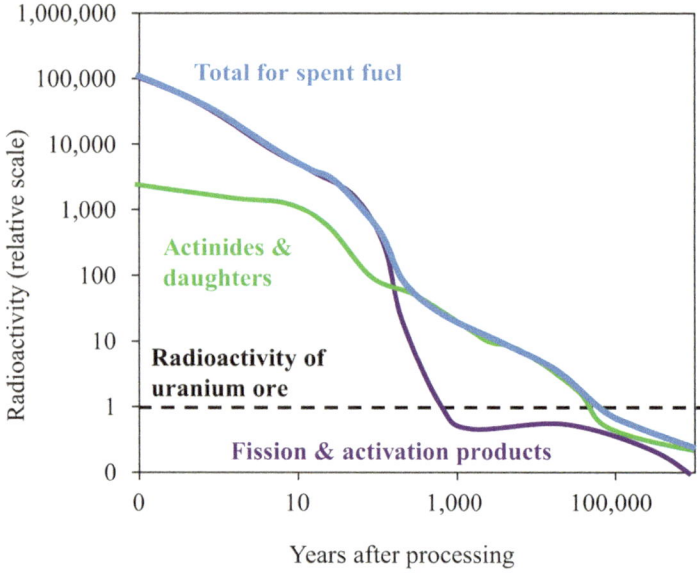

Fig. 3.2 A graph that shows the time until spent nuclear fuel radioactively decays to acceptable levels

would remain tracked and marked. Would we develop better ways to manage them with future technology? Technological change is accelerating. The task of predicting how this will go, many centuries out, is impossible. We just do not know and therein lies the core problem.

An important question is, can the waste ever come back up from a deep geological repository? Edwards makes a good point that you can't put waste into an undisturbed geological location because opening it up to put the waste in disturbs the geology! Once the repository is created, there is now a pathway to the surface that did not previously exist.

Further, he notes that nuclear waste is active. It generates increasing heat over time, with maximum heat at 4000 years and does not return to normal for 50,000 years. The radiation also generates ions that are chemically active (they led to the fire in the low-level depository in New Mexico mentioned above). Theoretically, even worse outcomes might be possible, including an accidental criticality (Gordon 2023).

Present considerations for preventing adverse outcomes revolve around the best way to encase high-level nuclear waste so as to contain it for very long time periods. A longstanding, approach is called vitrification, in which the waste is embedded in glass. This approach has appeal because glass is resistant to deterioration for timeframes comparable to the time it takes nuclear waste to reach levels of radiation similar to uranium ore. Ceramics are another option that have comparable persistence. As an example, ceramic artifacts, such as pieces of bowls, which were made thousands of years ago, can be recovered at archeological sites today.

A final, very worrisome, concern is that the waste contains, as noted above, plutonium. Plutonium can and is frequently used to make atomic bombs. It also has a half-life of 24,000 years so it will be present in the waste in substantial concentrations for tens of thousands of years. A present-day barrier to using this plutonium is that its extraction from the rest of the waste is extremely dangerous.

The other highly radioactive elements from which the plutonium must be separated are the source of radiation that could kill a person quickly (plutonium itself gives off only small amounts of radiation). With today's technology, one needs a robotically controlled facility called a reprocessing plant to isolate plutonium from nuclear waste. This is how Pakistan obtained plutonium for their first atomic bomb, isolating it from waste produced by a reactor given to them by Canada. Will it be easier or safer to extract plutonium in a thousand years? That seems possible, but no one knows.

If Finland, with which we started this chapter, is an example of success in terms of negotiating with the population adjacent to their nuclear waste repository, Taiwan's approach decades ago is a cautionary tale of the pitfalls of using deceit. In the 1970s, Taiwan's Atomic Energy Commission chose Orchid Island for their "temporary" storage facility for mid- and low-level nuclear waste.

As with other undesirable and potentially hazardous nuclear facilities, it appears the site was chosen because of the low population density and the low literacy level of the indigenous Yami people who inhabit the island. Although a recent New York Times story frames this as, "No one bothered to inform the residents", earlier documents suggest that the Taiwanese government was being deliberately deceptive (Qin et al., 2023). The Yami district commissioner was told at the time that the facility would be a fish cannery.

Eventually, the Yami realized what was happening and, "[a]round the Chinese New Year season in 1991, the Yami people rose up in protests which caught the attention of the media and public in all of Taiwan. Led by Kuo JIan-ping, a Yami Presbyterian missionary, and with the support of anti-nuclear groups in Taiwan like the Taiwan Environmental Protection Union and the Green Association, the Yami anti-nuclear group held demonstrations on Orchid Island and in Taipei (Fig. 3.3), where they carried a protest letter straight to the Taiwan Power Company." (Marsh et al., 1993).

The experience of the Yami, who continue to live with tens of thousands of containers of nuclear waste, despite new deliveries being halted by their protests, is reminiscent of the many cases of deceit and imposition of risk on indigenous communities in the United States and elsewhere around the

Fig. 3.3 Biking Projection against nuclear power plant (NPP)'s Construction in Taiwan (Greenpeace, 2021)

world. Whatever the solution to disposal of waste from nuclear power plants, let's agree that further exploitation of indigenous lands should be off limits.

Summary Points

1. Radioactive waste must be contained and managed for a duration longer than the age of civilization.
2. In the U.S., there is currently no plan for long-term storage of waste generated by civilian nuclear power generation.
3. As with mining and processing of uranium, the disposal of high-level nuclear waste has also affected indigenous communities.

References

Benjamin, M. (2023). Focus: France seeks strategy as nuclear waste site risks saturation point. *Reuters* https://www.reuters.com/business/environment/france-seeks-strategy-nuclear-waste-site-risks-saturation-point-2023-02-03/

Corkhill, C. L., & Hyatt, N. C. (2018). Nuclear waste management. *Research Gate*. https://doi.org/10.1088/978-0-7503-1638-5

El-Showk, S. (2022). Final resting place. *Science (American Association for the Advancement of Science), 375*(6583), 806–810. https://doi.org/10.1126/science.ada1392

Gordon, E. (Producer), & Gordon, E. (Director). (2023). *Nuclear waste the questions multiply.* [Video/DVD] YouTube. https://www.youtube.com/watch?v=uEG3pS0bUvk

Greenpeace. (2021). *No to nuclear.* Taiwan: Taipei times. https://www.taipeitimes.com/News/taiwan/archives/2021/08/29/2003763432.

Marsh, D. R., Lin, E., & Lin, P. (1993). *Orchid Island: Taiwan's nuclear dumpsite.* World Information Service Energy. https://www.wiseinternational.org/nuclear-monitor/387-388/orchid-island-taiwans-nuclear-dumpsite

NMWO. (2023). *What other countries are doing.* www.nwmo.ca. Retrieved July 24,2023, from https://www.nwmo.ca/Canadas-Plan/What-Other-Countries-Are-Doing

Qin, A., Chien, A. C., & Fei, L. Y. (2023). The nuclear dump that created a generation of indigenous activists. *The New York Times.* https://www.nytimes.com/2023/01/05/world/asia/lanyu-taiwan-nuclear-waste.html

Chapter 4
Nuclear Proliferation

On March 29, 1976, former US Government officials who had worked on nuclear weapons released statements jointly under the headline, "The Peaceful Atom Goes to War". The officials were:

Dr. George Kistiakowsky, Head of Explosives Division for the Manhattan Project and Special Assistant to President Eisenhower;

Dr. Theodore Taylor who had been a nuclear weapons designer at Los Alamos Laboratories and Deputy Director of the Defense Department's Atomic Support Agency;

Herbert Scoville, formerly Technical Director of the Armed Forces Special Weapons Project and Head of Scientific Intelligence at the CIA;

Dr. George Rathjens, previously Director of Weapons System Evaluation and former Chief Scientist, Advanced Research Projects Agency, Department of Defense; and

Dr. Bernard Feld, Assistant Leader of the Critical Assembly Group, WW II Atom Bomb Project, Former Secretary General, Pugwash International Scientific Conferences, Vice-President of the American Academy of Science Editor in Chief, Bulletin of the Atomic Scientists.

Scoville went first, laying out the overall theme of their mission:

> The four of us are assembled here today at Princeton in the office which was being used by Professor Albert Einstein when the awesome potentialities of a nuclear explosion were first recognized. As a result of discussions in this very office, Einstein wrote to President Roosevelt urging a programme to ensure that this dangerous weapon did not fall into Nazi hands. This was the genesis of the atomic bomb.

Doug Brugge is the primary author of this chapter.

© The Author(s) 2024

D. Brugge, A. Datesman, *Dirty Secrets of Nuclear Power in an Era of Climate Change*, https://doi.org/10.1007/978-3-031-59595-0_4

Now some thirty years later we are gathered here because we are concerned that these weapons will soon fall into many hands in many corners of the world – into the hands of unstable national governments, aggressive military cliques or irresponsible terrorist groups, with incalculable consequences for us all. *This danger is the direct result of the uncontrolled growth of the nuclear power industry, which is making widely available the materials needed for such weapons*" (CCNR, 1976)

What is remarkable in that quote is the unambiguous link to nuclear power that was evident to this highly knowledgeable expert in the field of nuclear technology and his colleagues over 40 years ago. Despite the certainty of this connection through the interceding decades, there continues today to be attempts to obfuscate and blur the issue in the public eye.

Writing in the New York Times, Joshua S. Goldstein and Staffan A. Qvist, authors of "A Bright Future: How Some Countries Have Solved Climate Change and the Rest Can Follow" along with Steven Pinker, professor of psychology at Harvard University, claim nuclear power has, "not contributed to weapons proliferation, thanks to robust international controls: 24 countries have nuclear power but not weapons, while Israel and North Korea have nuclear weapons but not power" (Goldstein et al., 2019).

The authors make this claim by including 16 countries that developed nuclear power while under the umbrella of either NATO or the Soviet Union. They also claim that North Korea and Israel developed nuclear weapons without nuclear power, but that is also deceptive because both have nuclear reactors, just not commercial use of nuclear power to generate electricity.

With the five major nuclear powers (US, UK, Russia, France and China) that leaves a number of countries that either were in the past, are or could become threats of gaining nuclear weapons. These countries are: South Africa, Argentina, Brazil, Bangladesh, Egypt, Iran, and the United Arab Emirates. It also leaves aside India and Pakistan, who developed nuclear weapons outside the non-proliferation agreement (Fig. 4.1).

The primary challenge to making a nuclear weapon is obtaining fissionable material. There are two types of fissionable material that can be used, either highly enriched uranium or plutonium. Natural uranium comes in two "isotopes" that have different molecular weights of 235 and 238. Only uranium 235 can be used in a nuclear reactor or bomb.

Uranium 235 is present in only tiny quantities in natural uranium. There are several industrial processes that enrich uranium 235. These are gaseous diffusion, gas centrifugation and use of lasers. Low level enrichment of uranium to 3.5–5.0% uranium 235 is necessary to produce fuel for nuclear

a Country position on nuclear weapons, 2023

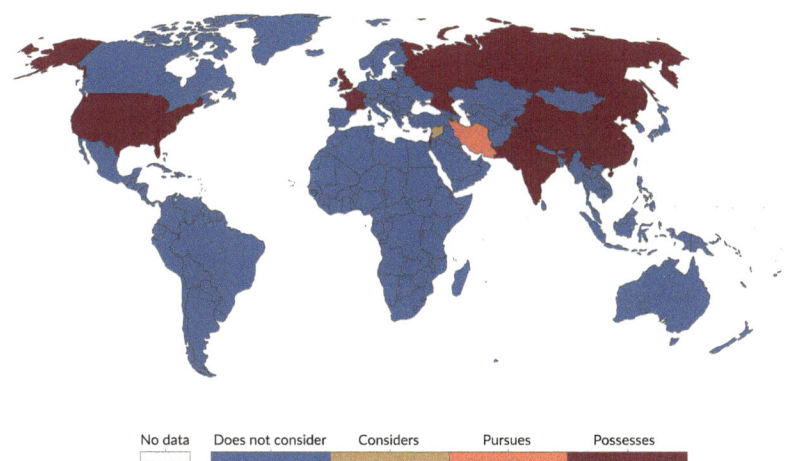

| No data | Does not consider | Considers | Pursues | Possesses |

Data source: Bleek (2017); Nuclear Threat Initiative (2024)   OurWorldInData.org/nuclear-weapons | CC BY
Note: The Chart tab uses numeric values, ranging from 0 for not considering nuclear weapons, to 3 for possessing them.

Source: Our World in Data

b

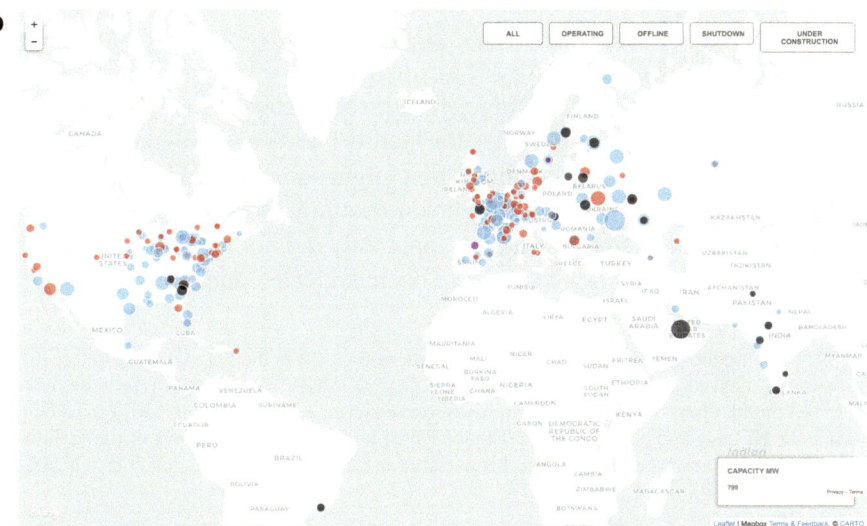

Source: Carbon Brief

Fig. 4.1 Maps of countries (**a**) possessing, or having sought, nuclear weapons and (**b**) the location of nuclear power plants largely coincide, with the notable exception of Japan and Germany, both of which were demilitarized after World War II and effectively operate under the protection of US nuclear weapons. ((**a**) Source: Our World in Data (n.d.), (**b**) Source: Carbon Brief (2016))

reactors. Greater than 90% uranium 235, or highly enriched uranium, is required for use in bombs.

The same processes can be used to enrich to levels for use in nuclear power plants or for production of bomb grade material. Therein lies a problem with extending nuclear technology for peaceful, power generating purposes to countries that might want to make weapons. They can claim, as has Iran in recent years, that they are enriching uranium for peaceful purposes even if they are probably aiming to enrich to a higher grade for use in weapons (Crowley et al., 2023).

Most people assume uranium and plutonium are hazardous. However, their fear is often focused on the health consequences of exposure to radiation, which is not the main concern they should have, as both elements release only low levels of radiation. Rather, the serious risk is that very small quantities of enriched uranium and even smaller amounts of plutonium, can initiate a nuclear fission chain reaction. An unintentional event of this nature is called a criticality.

An historical example that is not well known outside of those with deep knowledge of nuclear events in the US serves to illustrate this danger. Karl Z Morgan was a physicist and a founder of the field of health physics. Toward the end of his life he became critical of nuclear power and nuclear weapons. In his autobiography, "The Angry Genie: One Man's Walk Through the Nuclear Age", he retells the story of a criticality at the Union Carbide Nuclear Company, Y-12 Plant at Oak Ridge, Tennessee (Crowley et al., 2023).

Because of the risk of a criticality, the facility had a prohibition against bringing in even relatively small containers that might somehow end up with enough enriched uranium in them to start a chain reaction. While most of the people working there had been trained, they apparently did not think to train the janitor, a Black man, who might also have been neglected due to his race.

Morgan writes:

> One morning the janitor commenced his early morning tasks in the Y-12 building before the operators arrived. Annoyed that a puddle of dirty yellowish solution had repeatedly collected on the floor, he 'solved' the problem. He retrieved a 55-gallon rain barrel from outside the building and placed it under the pipe where it would catch the slowly dripping fluid. Day after day and week after week this barrel remained in an inconspicuous place behind some machinery.

Until finally, in Morgan's words:

> I was in my office at X-10 that morning when the phone rang. I picked up the receiver to hear someone shouting, 'We have a criticality accident at Y-12 and thousands of employees are evacuating the plant!'

> I reached for my emergency kit and rushed for the door. My assistant, Hubert Yockey, grabbed his kit as well, and we ran out to a company car. I drove the ten-mile distance over a rough sandy road to Y-12 in eight minutes. Hundreds of persons milled outside the gate.

Only our car was permitted past the guards and allowed to enter the area where minutes before thousands of people had been at work. When Yockey and I entered the windowless building that contained the problem, darkness engulfed us. I muttered to myself, 'My king-dom for a flashlight.' …. A faint light shone from a battery operated emergency lamp in the far end of the building, and we 'homed in' on the life-threatening barrel as best we could. Unable to read the scales on the Geiger counter, we could hear the clicks sounding faster and faster as we approached the far end of the building.

Each time the Geiger counter needle banged the end of the scale, we changed to a higher scale as we approached the radioactive source. Most important, we listened for clicks on our Hurst neutron dosimeter. Fortunately, we heard none. The presence of neutrons would mean a life-threatening critical assembly still existed.

The clicks from our Geiger counter saturated or ran together on the highest scale, so the counter stopped clicking. …. We ran from the building." After they reported out, a team "put on protective clothing and masks, rushed into the building, and poured into the barrel a high concentration of borax, which absorbs neutrons and "kills" any possibility of a criti-cal assembly of the fluid.

Blood samples taken promptly afterward revealed that many employees had received "impermissibly high neutron and associated gamma dose[s]." "Five of the Y-12 workers experienced radiation sickness and [loss of hair]. Those who received [high doses] of radiation experienced some [bleeding] …. Even the individual who received [a lower dose] showed some symptoms of radiation injury." (Crowley et al., 2023).

The point is that fissionable materials are very dangerous, in even tiny amounts.

The other source of fissionable material is plutonium. However, the source of plutonium is different than fissionable uranium. Plutonium exists in only trace quantities in natural uranium deposits, but it is produced by fission in nuclear reactors. To obtain enough for use in nuclear power or weapons, it must be chemically removed from the other highly radioactive byproducts of nuclear fission.

Extracting and purifying plutonium from high level nuclear reactor waste is a hazardous industrial operation (Fig. 4.2), but a country hoping to obtain nuclear weapons capability need not undertake the process itself if it can find another country willing to sell plutonium. A telling example is India. In 1974, India used plutonium from a Canadian reactor to build and detonate its first atomic bomb in an underground test.

Thus, the extension of nuclear weapons status to India was based on political and economic decisions. Obtaining enriched uranium or plutonium remains the largest barrier to building a nuclear explosive. The rest of the design and construction consists of straightforward engineering issues. Less than 5 kilograms of plutonium is enough for one bomb.

It is concerning that there may be 1000 tons of plutonium produced, stored or used in weapons today (Wikipedia, 2023b). While there is tight security, it would take release, misplacement, or sale of a tiny amount to give

Fig. 4.2 A reprocessing plant that extracts plutonium from high level nuclear waste from nuclear reactors. The resulting plutonium can be used in both nuclear power plants and nuclear weapons. (Source: Aerial view Sellafield, Cumbria (2020), © Simon Ledingham: Geography Britain and Ireland)

a non-nuclear country enough to gain nuclear weapons. It's worth mentioning how much Pu and U-235 the US has lost, called, euphemistically, Material Unaccounted For (MUF). In 2012, the amount was 6 tons.

Canada is considered a peaceful country, however the potential for Canada to contribute to nuclear proliferation remains. Dr. Gordon Edwards, President of the Canadian Coalition on Nuclear Responsibility, has taken an active role opposing Canadian proposals to develop a nuclear reactor that depends on producing and extracting plutonium as fuel (called "breeder" reactors because they increase plutonium). The intention is to sell those reactors to other countries around the world, spreading access to the very technology required to be able to produce nuclear weapons.

A Canadian House of Commons committee recommended that the government "work with international and scientific partners to examine nuclear waste reprocessing and its implications for waste management and [nuclear weapons] proliferation vulnerability." The recommendation followed on a $50.5 million grant to the Moltex corporation awarded in March 2021 to "develop a plutonium reprocessing facility at the Point Lepreau nuclear site on the Bay of Fundy." (CRED-NB, 2023).

Dr. Edwards said in a press release, "By supporting the implementation of reprocessing technology intended for export, in connection with a plutonium-fueled nuclear reactor, without regard for the weapons implications, Canada may be once again spreading the bomb abroad," (CRED-NB, 2023).

When India achieved nuclear weapons status, it created the pressure for Pakistan, its arch enemy, to follow. That did not happen immediately though as the path to a nuclear weaponized Pakistan was convoluted and involved technology transfer from the US and China. According to the New York Times, China gave Pakistan the design for a nuclear weapon as well as highly enriched uranium (Weiner, 1998).

The US also helped Pakistan based on a geopolitical calculation that Pakistan was an enemy of India and India was closer to the Soviet Union, the strategic rival of the US. Following this logic, the US provided Pakistan with its first research nuclear reactor, giving them technologic skills necessary, but not sufficient, for building weapons. During the time when the US was backing Muslim rebels against the Soviet aligned Afghan government, the US, in the words of the NYTs, "turned a blind eye" to Pakistan's nuclear weapons program (Weiner, 1998).

Unlike Pakistan, for whom obtaining a nuclear reactor played a small part in developing nuclear weapons, building a nuclear reactor was central to the path by which Israel became a nuclear power. While Israel neither confirms nor denies its nuclear capacity, it is well established that they have hundreds of nuclear warheads, although the exact number and nature – tactical vs. full sized – and their delivery mechanisms remain unclear.

The inability to separate "Atoms for Peace", the Eisenhower program to spread nuclear power without weaponization, is inherent in the Israeli nuclear program. Israel eagerly signed onto Atoms for Peace, while secretly using peaceful intent as a cover for developing nuclear weapons, an approach repeated by other would-be nuclear players in later years.

Under the guise of peaceful nuclear power to be used to desalinate seawater that would irrigate the desert, France helped Israel build its first nuclear reactor at Dimona (Fig. 4.3). For our purposes, the point is that this reactor was used to generate plutonium, separated from the rest of the waste at a reprocessing center and then used in nuclear weapons. To be clear, plutonium extracted from fuel rods in the reactor was critical, but it was not sufficient by itself. Additional technology and skills were also obtained, including heavy water secretly shipped from the UK through Norway and yellow cake, concentrated uranium from ore, provided by Argentina (Burr & Cohen, n.d.; Wikipedia, 2023a).

As with other nuclear powers and those seeking to join the club, Israel sought nuclear weapons for strategic, geopolitical purposes, primarily to have a deterrent to attacks from neighboring states. That motivation remains compelling to many other countries, including others in the Middle East. Given its own trajectory, Israel understands better than most how nuclear power technology is a step toward having a bomb.

Fig. 4.3 A picture taken on March 8, 2014, shows a partial view of the Dimona. (News Photo – Getty Images)

This is precisely why Israel destroyed nuclear reactors in Syria and Iraq and, together with US help, that damaged nuclear technology that Iran, not surprisingly, argues is for peaceful purposes (Burr & Battle, 2021; Farrel, 2018; News Wires, 2021; The Iran Primer, 2021). Inspections in Iran have centered around how highly they are enriching uranium, is it low grade for medical purposes, or high grade for bombs? Seeing through the veil of secrecy to discern which it is, is not so easy. Recently though, it has become apparent that Iran is very close to having the enriched uranium it needs for a weapon.

Many other countries started down the nuclear path before abandoning it. These included South Korea, Taiwan, Argentina, and Brazil. Only South Africa succeeded in developing nuclear weapons before abandoning them. At the core of the South Africa program, billed as developing "peaceful nuclear explosives", was, as elsewhere, a nuclear reactor, in this case provided by the US (Albright, 2001).

Some have claimed that thorium reactors are a solution because this type of reactor is not a threat to proliferation. This is incorrect because thorium is not actually a nuclear fuel since it cannot sustain a nuclear chain reaction. When thorium is mixed with plutonium, our old friend, the resulting "mixed fuel" can sustain a chain reaction thanks to the fissile plutonium. The resulting neutron bombardment converts a portion of the inert thorium into fissile uranium-233. (Gordon, 2023).

It should be particularly disturbing to anyone who supports non-proliferation that North Korea, a low-income, economically undeveloped, secretive and isolated country was able to obtain nuclear weapons. North Korea obtained its first, small, research grade reactor from the Soviet Union and subsequently had access to nuclear technology from Pakistan. While the country is shrouded in secrecy, at least some of its weapons use plutonium reprocessed from its nuclear reactors (NTI, 2021).

Today, other than North Korea and Iran, active attempts to circumvent non-proliferation are rare. Perhaps that reflects some success at convincing aspiring nations to forgo nuclear weapons or possibly some countries have begun to see the downsides of being a nuclear power. Whatever the reason, there is no certainty that the lull will continue. One wonders whether the possibility of extending nuclear power capacity to Saudi Arabia, an apparent component of negotiations underway as this is written, might open the door to a second nuclear power in the Middle East. (Murphy et al., 2023; Wilkins Brett, 2023).

It seems that possessing nuclear weapons remains a powerful incentive, especially in the Middle East. Nuclear power is a good cover for obtaining a start on the technology needed to build nuclear weapons. Do we want more and more states, even if they accumulate gradually, to possess these weapons? How many more pairs of enemies, like India and Pakistan, do we want to have staring each other down with nuclear weapons? Doesn't the risk of nuclear war increase the more enemies the world has in poses of mutually assured destruction?

Summary Points

1. Nuclear power cannot be disentangled from the potential to develop nuclear weapons.
2. Pakistan and Israel utilized nuclear reactors to develop nuclear weapons.
3. Because only a small amount is required to create a weapon, plutonium is a particular concern.

References

Aerial view Sellafield, Cumbria. (2020). https://www.geograph.org.uk/photo/50827
Albright, D. (2001). Institute for Science and International Security. South Africa's nuclear weapons Program. http://web.mit.edu/SSP/seminars/wed_archives01spring/albright.htm

Burr, W., & Battle, J. (2021). Israeli attack on Iraq's Osirak 1981: Setback or impetus for nuclear weapons? *Iran Iraq nuclear Vault from National Security Archive: 767.* https://nsarchive.gwu.edu/briefing-book/iraq-nuclear-vault/2021-06-07/osirak-israels-strike-iraqs-nuclear-reactor-40-years-later

Burr, W., & Cohen, A. (n.d.). *Kennedy, Dimona and the nuclear proliferation problem: 1961–1962.* Wilson Center https://www.wilsoncenter.org/publication/kennedy-dimona-and-the-nuclear-proliferation-problem-1961-1962

Carbon Brief. (2016). *Mapped: The world and nuclear power plants.* Carbon Brief, https://www.carbonbrief.org/mapped-the-worlds-nuclear-power-plants/

CCNR. (1976). *The threat of nuclear war.* Canadian Coalition for Nuclear Responsibility (CCNR). Granada television ltd, Manchester M60 9EA36 Golden Square, London W1R 4AH: Granada Independent Television/ the British Independent Television Network. https://ccnr.org/Peaceful_Atom.html

CRED-NB. (2023). *Spreading the Bomb – Will Ottawa revisit Canada's support for plutonium reprocessing?* Coalition for Responsible Energy Development in New Brunswick (CRED-NB). https://crednb.ca/2023/02/22/spreading-the-bomb-will-ottawa-revisit-canadas-support-for-plutonium-reprocessing/

Crowley, M., Fassihi, F., & Bergman, R. (2023). Hoping to avert nuclear crisis, U.S. Seeks Informal Agreement with Iran. *The New York Times.* https://www.nytimes.com/2023/06/14/us/politics/biden-iran-nuclear-program.html?smid=nytcore-ios-share&referringSource=articleShare

Farrel, S. (2018). *Israel admits bombing suspected Syrian nuclear reactor in 2007, warns Iran.* Reuters. https://www.reuters.com/article/us-israel-syria-nuclear-idUSKBN1GX09K

Goldstein, J. S., Qvist, S. A., & Pinker, S. (2019). Nuclear power can save the world. *The New York Times* https://www.nytimes.com/2019/04/06/opinion/sunday/climate-change-nuclear-power.html

Gordon, E. (2023). *Nuclear waste the questions multiply.* YouTube. https://www.youtube.com/watch?v=uEG3pS0bUvk

Murphy S. C, Hollen Van, C., Durbin J. R., Welch, P., Schatz, B., Carper, R. T., Duckworth, T., Murray, P., Baldwin, T., Sanders, B., Fetterman, J., Warren, E., Ossoff, J., Shaheen, J., Kaine, T., Merkley A. J., Warnock, R., Markey, J. E., & Lujan Ray, B. (2023). In The Honorable Joseph R. Biden, Jr. President of the United States (Ed.), *U.S.-backed efforts to facilitate the normalization of relations between Saudi Arabia and Israel.* https://www.murphy.senate.gov/imo/media/doc/saudi-israel_normalization_letter.pdf

News Wires. (2021). *Israel strikes targets in Syria after missile lands near nuclear reactor.* France 24. https://www.france24.com/en/middle-east/20210422-israel-strikes-targets-in-syria-after-missile-lands-near-nuclear-reactor

NTI. (2021). *North Korea nuclear overview.* The Nuclear Threat Initiative (NTI). Retrieved August 4, 2023, from https://www.nti.org/analysis/articles/north-korea-nuclear/

Our World in Data. (n.d.). Country position on nuclear weapon. *Nuclear Weapons.* Our World in Data. Retrieved August 8, 2023, from https://ourworldindata.org/nuclear-weapons

The Iran Primer. (2021). *Israeli Sabotage of Iran's Nuclear Program.* www.iranprimer.usip.org. Retrieved August 4, 2023, from https://iranprimer.usip.org/blog/2021/apr/12/israeli-sabotage-iran%E2%80%99s-nuclear-program

Weiner, T. (1998). Nuclear anxiety: The know-how; U.S. and China helped Pakistan build its bomb. *The New York Times.* https://www.nytimes.com/1998/06/01/world/nuclear-anxiety-the-know-how-us-and-china-helped-pakistan-build-its-bomb.html

Wikipedia. (2023a). *Nuclear weapons and Israel.* www.wikipedia.org. Retrieved August 4, 2023, from https://en.wikipedia.org/wiki/Nuclear_weapons_and_Israel

Wikipedia. (2023b). *Plutonium.* www.wikipedia.org. Retrieved August 4, 2023, from https://en.wikipedia.org/wiki/Plutonium

Wilkins Brett. (2023). *Senators worry about Saudi nuclear arms plans.* Beyond Nuclear International https://beyondnuclearinternational.org/2023/10/08/senators-worry-about-saudi-nuclear-arms-plans/

Chapter 5
Societal Burdens Imposed by Nuclear Accidents

Nuclear power stations are like stars that shine all day long! We shall sow them all over the land. They are perfectly safe! (Medvedev, 1991)
— Academician M.A. Stryrikovich, Soviet power engineer

The nuclear meltdown at Chernobyl this month 20 years ago, even more than my launch of perestroika, was perhaps the real cause of the collapse of the Soviet Union five years later. Indeed, the Chernobyl catastrophe was an historic turning point: there was the era before the disaster, and there is the very different era that has followed … Chernobyl opened my eyes like nothing else: it showed the horrible consequences of nuclear power, even when it is used for non-military purposes. (Gorbachev, 2006)
— Mikhail Gorbachev, leader of the Soviet Union in 1986

The Canadian Meltdown

Although the technology was invented in the United States, the first meltdown of a nuclear reactor occurred in Ontario, Canada, when the NRX reactor at the Chalk River Laboratories suffered a serious accident in December 1952. The NRX reactor was a research facility, small by the standards of a modern commercial nuclear power station. In addition to radioactive gases that may have been vented to the atmosphere in the absence of monitoring, the accident is believed to have released 10,000 Curies of radioactivity contained within 1.2 million gallons of contaminated water that flooded the basement of the reactor building.

Aaron Datesman is the primary author of this chapter.

© The Author(s) 2024

D. Brugge, A. Datesman, *Dirty Secrets of Nuclear Power in an Era of Climate Change*, https://doi.org/10.1007/978-3-031-59595-0_5

The Curie (Ci), named in honor of Marie Curie, is a unit of activity indicating a quantity of radioactive material. One Curie represents the number of disintegrations that occur per second in one gram of radium. It is
a large unit: one Curie indicates an activity of 37 billion disintegrations
per second. Often the Becquerel (Bq), representing just one decay per
second, is a more useful description. There is no direct, universal conversion between activity and dose. Each situation involving exposure to
ionizing radiation requires its own careful description and analysis.

It required the efforts of more than one thousand persons to clean up the
damaged reactor, which after a few years was placed back into operation. A
team from the Naval Reactors program in the U.S. was dispatched to lead
the repair effort. The person in charge of that unit of twenty-three individuals was a 28-year-old lieutenant from Georgia named James E. Carter. The
thirty-ninth president of the United States is pictured in his dress whites in
Fig. 5.1. President Carter shared the following recollection with a Canadian
journalist in 2011:

> It was the early 1950s ... I had only seconds that I could be in the reactor myself. We all
> went out on the tennis court, and they had an exact duplicate of the reactor on the tennis
> court. We would run out there with our wrenches and we'd check off so many bolts and nuts
> and they'd put them back on... And finally when we went down into the reactor itself, which
> was extremely radioactive, then we would dash in there as quickly as we could and take off

Fig. 5.1 (left) Lieutenant James Carter, who served with the Naval Reactors Branch of the
U.S. Atomic Energy Commission, headed by then-Captain Hyman Rickover. (right) President
Jimmy Carter visiting the stricken Three Mile Island Unit 2 nuclear power station, in April 1979

as many bolts as we could, the same bolts we had just been practicing on. Each time our men managed to remove a bolt or fitting from the core, the equivalent piece was removed on the mock-up. (Milnes, 2011)

Because the environment inside the NRX reactor building was so dangerous, individuals were permitted to enter in shifts lasting only 90 s. Even wearing protective gear, crew members acquired a dose equivalent to a year's worth of permitted exposure during each brief shift. The total dose Lt. Carter received exceeded limits considered acceptable today by a factor of about one thousand. He was told it was likely that he would never have children. (Thankfully, the prediction was not correct; President and Rosalynn Carter had four children.) Lt. Carter's urine was radioactive for 6 months after the accident.

Twenty-seven years later, President Jimmy Carter visited Central Pennsylvania in the wake of a serious accident at the Three Mile Island Unit 2 nuclear power station in Middletown, PA, near the state capital of Harrisburg. The accident began early in the morning on Wednesday, March 28, 1979. Carter is pictured in the TMI-2 control room, on Sunday, April 1, 1979, in the photograph on the right in Fig. 5.1.

All Technologies Are Accident-Plagued

Accidents are how engineers learn. There should be nothing surprising about this simple idea. For instance, if one desires to learn the fracture strength of a material, it is necessary take a piece and break it. "Engineering is an art form that makes use of scientific principles," wrote journalist Ira Rosen and engineer Mike Gray, "and this marriage confuses a lot of people. We tend to think of engineering itself as a science, but it is nothing more than advanced carpentry. The practitioners learn by doing." (Gray & Rosen, 1982) Their correct insight appears in the introduction to their book about the Three Mile Island accident, titled The Warning.

Accidents happen, full stop: any other position is fantasy on the level exhibited by the Academician in the first quotation. Despite public-facing assurances of safety and competence, in fact the authorities understand the reality of the situation. For instance, the Nuclear Regulatory Commission in 1985 asserted as a goal that there should be not more than one incident of core damage per 10,000 years of reactor operation. There have been at least ten incidents of core damage over the history of nuclear power technology in Western nations.

Although one might argue over whether certain instances should be excluded on the basis of dual use (both military and civilian) or experimental design, it is inarguable that four commercial nuclear power stations designed by U.S. firms have melted down, with accompanying large releases of ionizing radiation. These are Three Mile Island (TMI) Unit 2, in Pennsylvania in 1979, and Fukushima Daiichi Units 1, 2, and 3, in Japan on March 11, 2011. Additionally, one commercial power reactor of Soviet design melted down[1] in Ukraine, on April 26, 1986. That facility was the Chernobyl Nuclear Power Plant Unit 4, a graphite-moderated RMBK reactor lacking a containment structure.

As of 2019, worldwide cumulative reactor operating experience exceeded 18,000 reactor-years. Therefore, considering only these five serious accidents, the frequency of meltdown events (much more severe than "core damage") accompanied by a serious radiological release has been one per 3600 reactor-years. The historical performance of nuclear power stations has not met NRC's stated goal for operational safety.

Common Threads

The nuclear disasters in Ukraine and in Japan were vast and complex events, fundamentally disruptive to the societies in which they occurred. For instance, as stated in the second quotation opening this chapter, the Chernobyl disaster may have been the most significant single factor contributing to the dissolution of the Soviet Union. No useful effort can be made in this space to provide a comprehensive description of either accident. The authors wish instead to examine the societal dynamics surrounding nuclear power plant accidents in the context of the Three Mile Island accident, with which they have a connection through ongoing scientific work. There are a handful of common threads: contamination of foodstuffs, long-distance dispersion, and government and corporate secrecy, along with displaced populations, and massive personnel requirements for cleanup and remediation.

The consensus position of the scientific establishment (as expressed by the National Academy of Sciences BEIR VII committee) is that there is no such thing as a safe dose of ionizing radiation. Every exposure carries with it the possibility of harm, in a manner that increases with the degree of exposure. Nevertheless, there exist exposures that are "allowable," including levels of contamination in drinking water and foodstuffs.[2] In the U.S., per-

---

[1] There is reasonably convincing evidence, based on seismology and isotopic analysis, that a low-yield nuclear explosion took place at Chernobyl Unit 4, in addition to the steam explosion that destroyed the reactor.

[2] Because at least the environmental component of background radiation cannot be avoided, the idea of enforcing reasonable regulation somewhere above levels of zero contamination seems very reasonable. For instance, what sense would it make to limit strontium-90 or iodine-131 to levels below the natural level of potassium-40 (K-40)? Bananas contain K-40 at a level of around 120 Bq/

mitted levels of radioactivity in foodstuffs are set by the Food and Drug Administration (FDA). The radioisotopes cesium-134 and cesium-137 are permitted at levels up to 1200 Bq/kg. The regulation limiting cesium radio-isotopes in the European Union is set at 600 Bq/kg. Meanwhile, according to the Food Sanitation Act in Japan established in 2012, the permitted level of radiocesium in general foodstuffs is much lower: only 100 Bq/kg. It should therefore be understood that permitted levels of contamination are a legal matter – in fact, a social and political determination. They are not scientific.

The accidents at Chernobyl and Fukushima were not localized events. As was also true during the era of nuclear weapons testing,[3] the deposition of fallout due to these accidents was worldwide. The international community first became aware of the Chernobyl disaster because radioactivity was noticed in Sweden; the mainstream academic journal *Health Physics* has a topic keyword "Turkish tea" due to contamination from the accident. Radiation from Fukushima likewise appeared on the West Coast of the United States within only a short time after the accident. Radioactive contamination was found in produce grown in California by the technical staff of the Department of Nuclear Engineering of the University of California at Berkeley.

Secrecy and distrust are a final common element. For instance, the New York Times reported the following about investigations of the Chernobyl disaster undertaken in 1991, the year the Soviet Union dissolved:

> The opportunity to ask questions, limited as it may be, has still allowed the Supreme Soviet's commission investigating Chernobyl to uncover two high-level secret government orders: one from 1987 classifying as secret any information on the extent of radiation contamination, and one from 1988 decreeing that no medical diagnosis may connect an illness with radiation exposure. (Barringer, 1991)

The nuclear power plant accidents at Chernobyl, Fukushima, and Three Mile Island were monumental events with vast consequences, for which governmental authority bore significant culpability. Information provided by these same governments should not necessarily be regarded as trustworthy, authoritative, or complete, even in the absence of other sources of reliable information. The reality that one's government cannot be fully trusted is one of the costly societal burdens imposed by nuclear power technology.

---

kg. Because the concentration of potassium in the human body is homeostatically controlled within a narrow range, the comparison is unfortunately false.

[3] According to UNSCEAR, global deposition of Cs-137 and I-131 due to weapons testing was 26 MCi and 18,200 MCi, respectively. The quantities greatly exceed the fallout due to reactor accidents, although dispersed over a far greater area.

Chernobyl and Fukushima

The two most significant radioisotopes released by nuclear power plant accidents are believed to be cesium-137 (Cs-137) and iodine-131 (I-131). According to an evaluation published by the OECD Nuclear Energy Agency, the Chernobyl accident released about 2 million Curies (2 MCi) of Cs-137, and 27 MCi of I-131, into the environment. The reactor's entire inventory of the radioactive noble gas xenon-133 was also released, although because it dispersed widely the impact of this release is generally neglected. About 3% of the fuel escaped as particles, amounting to as much as six tons of highly radioactive material.

The 49,000 residents of the town of Pripyat – an *atomgrad*, or atomic city, with special privileges and a high quality of life – were evacuated in haste on 1100 buses the day after the accident began. Although most believed they would return to their homes after a few days, in fact Pripyat was abandoned after the accident. Altogether about 200,000 people were relocated from contaminated areas after the Chernobyl disaster. The criterion for relocation was a level of Cs-137 exceeding 40 Curies per square kilometer. Spread uniformly, by this criterion the 2 MCi of radiocesium released by the accident would render an area of 18,000 square miles uninhabitable. The area is three-quarters of the size of the entire state of West Virginia. The actual off-limits area amounts to about 6000 square miles, about the size of the state of Connecticut.

The idea of allowable levels of contamination was an urgent matter for authorities in the Soviet Union in the wake of the Chernobyl meltdown, especially since the affected regions were important agricultural areas. For instance, meatpackers were given special instructions on how to process radioactive meat.

> The instructions ordered butchers to grade the meat by radioactivity. Packers were to grind up radioactive flesh and mix it with appropriate proportions of clean meat for sausage. The experts in accident logistics were thinking along the commonly understood belief that diffusion[4] was the solution. Spread the contaminated meat broadly so each person across the vast USSR unknowingly ingested their own small part of the tragedy. Preparing the goods for sale, the packers were told to "label the sausage as you normally would." (Brown, 2019)

Meanwhile, privileged sectors of the society, including KGB employees, took measures to ensure that they received foodstuffs free of radioactive

---

[4]This idea often goes by the catchier expression "Dilution is the solution to pollution." Unfortunately, diluting the exposure is another form of protraction. If the hypothesis of shot noise in radiobiological systems is correct, mixing contaminated meat with clean meat in this manner increased the equivalent dose to the population.

contamination. When radioactive meat turned up in Moscow, messages were sent to Kyiv demanding that such shipments should not reoccur.[5]

While striving to manage the disaster, it was necessary also for the authorities to learn about what was happening, and to try to understand it.

> Kyiv researchers kept a close eye on livestock in farms in the Narodychi and Chernobyl regions, areas that were heavily contaminated and easy to reach from Kyiv ... "In fact," the researchers summarized in 1988, "damage over a protracted period does not correspond with the [non-acute or low] dose, but looks like acute radiation symptoms." The researchers suggested the need to recalculate the method of extrapolating the effects of large doses to small doses. (Brown, 2019)

The idea that protracted exposures may be much more dangerous than commonly understood is a theme of other chapters in this book.

The "liquidation"[6] of the Chernobyl disaster, "was a task on a scale unprecedented in human history, and one for which no one in the USSR – or, indeed, anywhere else on earth – had ever bothered to prepare." (Higginbotham, 2019) At least seven hundred thousand people – many inadequately equipped, poorly trained, and unaware of the risks – contributed to liquidation efforts. One of the liquidators, a radiation biologist named Natalia Manzurova, worked in the exclusion zone for more than four years. She recorded the following memory of her experience, relating to the abandoned kindergarten pictured in Fig. 5.2.

> I was amazed by the luxury of that kindergarten when I visited it to look for furniture I could use in the new lab and office. There were Chinese rugs and different matching color schemes for curtains and bedspreads in each sleeping room and a sea of stored toys, visual aids and games. New bed linen, towels, aprons and white dressing gowns were neatly piled and hung up. Looking at the rows of children's slippers and photos of their owners on the wall, I wondered where they might be now and how they were doing...
>
> One time when I touched a table in the kindergarten, I felt a jolt of pain in my thumb. I had probably touched a 'hot particle', the same type of large radioactive particle that injured Chernobyl's first liquidators through inhalation and skin burns. It hurt immediately and my finger swelled, turned a blue-lilac color and later the skin peeled off. (Manzurova & Sullivan, 2006)

Because the GE boiling water reactors that melted down possessed containment vessels, and because most of the fallout from the disaster blew eastward, into the Pacific Ocean, the 2011 accident in Japan seems to have been less severe than Chernobyl. According to modeling performed by the Japanese Atomic Energy Agency, the releases from the Fukushima disaster

---

[5] The authors rely heavily upon the excellent book <u>Manual for Survival</u>, by M.I.T. historian of science Kate Brown. We made this choice because Prof. Brown did what an outstanding investigator should do: she spent years in the state archives of Ukraine and Belarus, examining primary documents. In many cases, she reported, she was the first person ever to have done so.

[6] The curious phrase "liquidation" was due to Mikhail Gorbachev, in his first televised address after the accident.

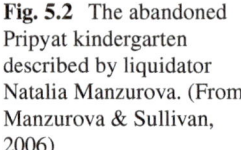

Fig. 5.2 The abandoned Pripyat kindergarten described by liquidator Natalia Manzurova. (From Manzurova & Sullivan, 2006)

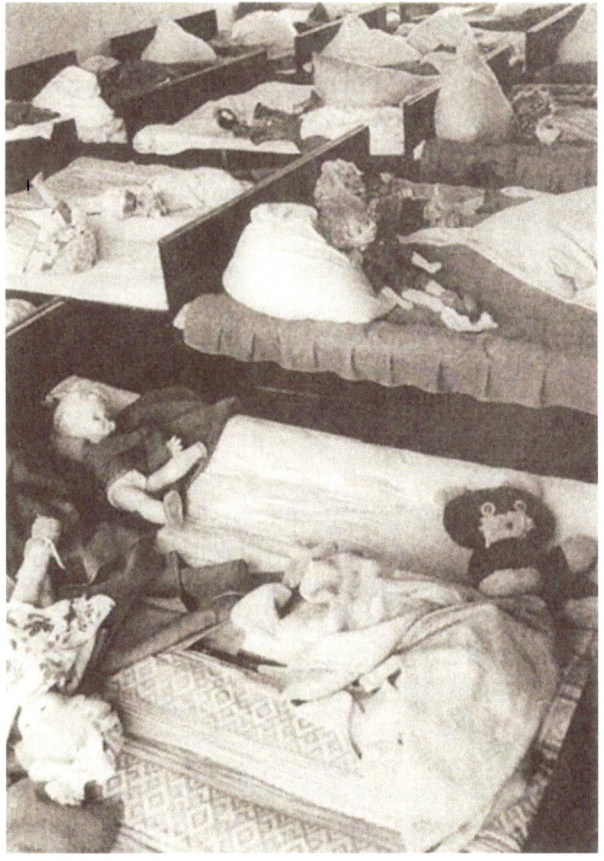

were about one-tenth as great as those due to Chernobyl: 0.3 MCi Cs-137, and 3.2 MCi I-131. Since only about 20% of the radioactivity released contaminated land, the exclusion zone near Fukushima has an area of only 143 square miles. Like the Chernobyl accident, however, the radioactive contamination released by the accident spread worldwide. The sensitive monitoring network operated by the Comprehensive Test Ban Treaty Organization measured fallout from the Fukushima accident even in the southern hemisphere within a month of the tsunami and triple meltdown.

> Nine days after the accident, the radioactive cloud had crossed Northern America. Three days later when a station in Iceland picked up radioactive materials, it was clear that the cloud had reached Europe ... As of 13 April 2011, radioactivity had spread to the southern hemisphere of the Asia-Pacific region and had been detected at stations located for example in Australia, Fiji, Malaysia and Papua New Guinea. (CTBTO, 2011)

The number of persons displaced from contaminated areas near Fukushima Daiichii in Japan is nearly 120,000. Although a liquidation effort at the scale of Chernobyl was not implemented in 2011, more than 77,000 persons worked in remediation efforts at Fukushima through the first five years after the accident. Human rights experts working for the United Nations have expressed concern over exploitation and coercion of those individuals, whose health may also have been negatively impacted. There were 46,000 individuals employed at Fukushima in 2016. Cleanup efforts continue as of 2023 and will continue for many additional years.

The idea that protracted low-level exposures may be more harmful than the authorities assert is consistent with the following observation from Japan in the wake of the accident at Fukushima Daiichi:

> One of Japan's most vocal physicians, 95-year-old Shuntaro Hida, charged in the summer of 2012 that people in Japan were already starting to develop symptoms of internal radiation poisoning, including fatigue, diarrhea, and hair loss, resulting from the ingestion and/or inhalation of radioisotopes. Dr. Hida is a native of Hiroshima. After the bombing there he treated patients exposed to the fallout ... Hida told the *Japan Times*: "I am worried because I received calls much earlier than I expected." (Nadesan, 2013)

A rigorous scientific test of the hypothesis, described more fully in a separate chapter, remains for the future.

An Impossible Battle Against Dust

As a young man, one of the authors (AD) for a time held a work-study job in a chemical engineering department. The work was involved with the topic of combustion aerosols. He remembers the safety protocol for a liquid mercury spill (one should sprinkle powdered sulfur on it), and the insight offered by the graduate student mentor. "The more you know about dust," the mentor said, "the less you want to breathe it."

While pure uranium is metallic, for use in a nuclear reactor the uranium is converted into a ceramic powder of uranium oxide, pressed into pellets about 3/8" in diameter and 5/8" long. The material is brittle, fracturing along grain boundaries due to thermal stress. The fracture stress decreases as fission products build up during operation of the reactor; that is, the integrity of the material degrades the longer the reactor operates. The fuel pellets (there may be as many as ten million of them in a reactor) are encased in long, slender fuel rods made of zirconium metal.[7] Zirconium alloys are chosen for this application due to their low neutron cross-section and excellent corrosion resistance. However, the material becomes brittle during reactor operation due to both corrosion and irradiation. Under certain circumstances,

---

[7] One of the biggest hurdles overcome by the Naval Reactors program on the road to developing the reactor for the Nautilus submarine, and subsequently commercial nuclear power, was the creation of an industrial supply chain for the production of pure zirconium metal.

it can even burn. The Chernobyl RMBK reactor furthermore utilized a graphite moderator, adding yet another brittle ceramic material to the mix.

The appearance of such terms as "ceramic", "brittle", and "fracture" (not to mention the idea that the cladding and moderator might burn) should evoke an important, though prosaic, concern: dust. The loss of containment accidents at Chernobyl and Fukushima released vast clouds of radioactive dust to the environment. One of the locations the radioactive dust created by the meltdown and explosion at Chernobyl Unit 4 wound up was in the fuzzy wool coats of sheep.

> In a chorus of voices, the women described the slowly dawning realization in the summer of 1986 that the distant nuclear accident had entered their lives ... By the end of May, many workers suffered mysterious nosebleeds. They complained of scratchy throats, nausea, and fatigue. Union records show that a couple of drivers, after helping out in the fields, sought medical treatment. In the sorting shop, the hay bales measured up to 30 µSv/hr. The wool workers did not know that picking up the most radioactive bales was like embracing an X-ray machine while it was turned on. (Brown, 2019)

The reactors at Fukushima Daiichi in Japan possessed containment structures that the Chernobyl RMBK reactor lacked. Therefore, for technological as well as for sociological, cultural, and political reasons, the situation in Japan is not the same as in Ukraine and Belarus. (It might be more correct to make a comparison between certain areas of Japan and areas of Europe heavily contaminated with Chernobyl fallout. That topic is outside the scope of this discussion.) There are similarities, however. Soil is another place where radioactive dust winds up. Particles of radioactive cesium bind chemically to small particles of clay.

The only possible remediation is to scrape off and bury contaminated soil. At the Interim Storage Facility between the towns of Okuma and Futaba near Fukushima Daiichi, remediation efforts have buried 14 million cubic meters of soil − enough radioactive soil to fill the Tokyo Dome eleven times. The government of Japan has committed to moving this enormous quantity of material again, to a final disposal site, before the year 2045. Even if the plan as described can be executed successfully, the local environment will not become clean, since the nearby forests will remain highly contaminated. The situation in Japan has been perhaps less urgent, but it is certainly similar to the emergency disposal of highly radioactive wool, meat, and animal hides that was necessary in the agricultural regions surrounding Chernobyl a few decades ago.

Tiny particles of radioactive dust contaminating a vast quantity of buried topsoil may be difficult to visualize, but the issue possesses more immediate impact if those particles threaten to accumulate in one's own airways and body. On March 15th, 2011, while the accident at Fukushima Daiichi was

Fig. 5.3 Radioactive materials trapped on a breathing mask worn by an individual in Tokyo, 220 km from the Fukushima Daiichi Nuclear Power Plant, in 2011. The left and right sides of the same mask are shown. (Image from Higaki, 2023)

still ongoing,[8] a scientist employed by the University of Tokyo wore a face mask while mostly outside for eighteen hours. At that time, a plume of radioactive material from Fukushima Daiichi Unit 2 passed over the Kanto area of Tokyo, about 130 miles away. It was possible to create an image of the radioactive material captured by the mask by simply placing a plate with a photographic emulsion on top of it, as shown in Fig. 5.3.

The mask shown in Fig. 5.3, placed in a sealed bag, was subsequently placed in storage. It was not forgotten, however. Using a refined technique, the investigator recently examined the mask again, looking for the presence of localizable radiocesium-bearing microparticles around one micron in size. Twenty-two particles[9] were found, with a combined activity of about 8 Becquerels. The result was published in the mainstream academic journal *Health Physics*.

The devil is in the details. A vast collection of hazardous particles too small to see, which end up essentially everywhere and may be recycled through the environment over and over, may constitute an important, hidden or neglected dirty truth about the risks of nuclear power. Radioactive dust poses a difficulty without a very sensible solution.

---

[8] Since the molten cores of Units 1–3 have never yet (as of 2023) been precisely located, it is the author's opinion that the accident in fact remains ongoing even today.

[9] The absorbed dose due to one of these cesium microparticles is of the order of 10 nGy. However, while the situation is not identical to the case of a uniformly dispersed radionuclide like potassium-40, the hypothesis of shot noise in radiobiological systems does apply. The author finds a dose rate from the inhaled microparticle on the order of 10 mSv/hr., significantly above the K-40 "noise floor". The reader should note that the hypothesis remains unconfirmed.

Central Pennsylvania, 1979

The accident at Three Mile Island Unit 2 was not as severe as later events at Chernobyl and Fukushima. For instance, because there was no melt-through[10] of the containment vessel, widespread contamination of foodstuffs with radiocesium did not occur. Nevertheless, accident consequences lie on a spectrum. The same set of concerns identified for the more severe loss-of-containment accidents also arose in Central Pennsylvania. Identifying common themes in the context of a less cataclysmic event is therefore a useful exercise for understanding the societal burdens imposed by nuclear power plant accidents.

Like cesium contamination in root vegetables in Ukraine, or tritium accumulating in seafood in the Pacific Ocean near Fukushima, contamination of milk by radioiodine was an important worry in Central Pennsylvania. The nationally famous Hershey's Chocolate factory lies in the region, which has long been known for the quality of its dairy products. Long-distance dispersion of contamination was also observed. The only location where xenon-133 released during the initial phase of the accident was directly measured was a laboratory operated by the NY State Department of Health in Albany, NY, 375 kilometers away from Three Mile Island. Cleanup required more than a decade and $2 billion (in 2022 dollars).

Secrecy and distrust, relating to both government authority and to the corporate operator of the nuclear power station, are the final components of the discussion. Both were prominent aspects of the accident at Three Mile Island and its aftermath. For instance, while a general evacuation order was never given, the governor of Pennsylvania did recommend that pregnant women and young children should evacuate from a limited area on Friday, March 30 (two days after the accident began). As many as 150,000 people left the area, very often in a state of considerable panic. Resident Bill Peters, pictured in Fig. 5.4, shared the following recollection of his own decision to evacuate:

> (Friday afternoon) while in the process of leaving, the Fairview Township police come down the road and he hollered, "Bill, get the hell inside! I mean it. Get inside. Don't breathe the air! Close your doors and windows!" So I waved to him, I said, "Yeah...keep going!" (Laughter) "I'm getting out of here! I'm not staying!" (Smith Katagiri, 1989)

The governor's decision came amidst conflicting information and the absence of clear guidance. The situation is recorded by the NRC historian:

---

[10] Because authorities for several years asserted (against evidence) that the fuel had not melted, they were slow to acknowledge that a melt-through very nearly did occur. Even today this basic fact about the accident is routinely denied or ignored.

Fig. 5.4 Bill Peters at his
home near the Three Mile
Island nuclear power
station, in 1986. Mr. Peters
evacuated from the area on
Friday, March 30, 1979,
two days after the start of
the accident. In this image,
in his left hand he holds an
ordinary dandelion leaf. In
his right hand, he displays
a mutated dandelion leaf,
harvested from his
property. Gigantism is
known to be one impact of
ionizing radiation exposure
on plant life

The central concern at the White House, as at the NRC and the governor's office, was the advisability and feasibility of evacuation. William Odom of the National Security Council staff informed Zbigniew Brzezinski on Saturday morning [March 31, three days after the accident began] that "a major population crisis relocation would probably occur "sometime today". Other federal officials urged that the White House seriously consider recommending that [Pennsylvania Governor] Thornburgh order an immediate evacuation. (Walker, 2004)

In the author's view, it is in this context that President Carter's visit to Middletown on April 1st should be understood. The Three Mile Island accident precipitated a national security crisis, exactly as the disasters at Fukushima and Chernobyl did. Nuclear power technology is so dangerous that its failure is a matter of national security. The potential for serious crisis is the cost the technology imposes on society.

This national security crisis was traumatic for the affected community. The situation is summarized well, though somewhat dryly, in a recent review published in the journal *Risks, Hazards, & Crisis in Public Policy*:

By 1981, the prevalence of major depression and/or generalized anxiety was estimated to be 29%, and half of mothers interviewed expressed concern that their children's health would be affected ... Some women, classified as depressed immediately after TMI, continued to be symptomatic for as long as a decade afterward ... in the decades following TMI (1979–1998), deaths from heart disease were 67.2% higher among women and 32.1% higher among men exposed to the lowest likely level of radiation (<8 mrem) within the TMI 5-mile radius when compared with surrounding communities ... (Wilson et al., 2022)

The terrible impact to the community was evident to outsiders on the ground at that time. One of the experts brought in to deal with the hydrogen bubble in the TMI reactor was a member of the NRC staff named Victor Stello. Stello was a native of Pennsylvania who had served in the Army, where he lost an eye, and began his career helping to develop a nuclear-powered airplane with Pratt and Whitney.

> Having resolved the bubble issue to his own satisfaction, Stello, who was "a good Catholic," decided to attend Sunday mass in Middletown. The service was sparsely attended, and Stello was surprised when the priest offered general absolution to the congregation. The rite was given in rare cases where ... large-scale loss of life seemed imminent. It was an emotional moment for the parishioners. "Everybody started crying, and I started crying," Stello recalled ... He returned from the church service in a highly emotional frame of mind and remarked unhappily to [NRC Public Affairs Officer] Joe Fouchard, "Look what we have done to these fine people!" (Walker, 2004)

Summary Points

1. Accidents at commercial nuclear power stations are national security incidents. The burden imposed by a societal crisis of this sort is vast, and likely immeasurable.
2. Governments and the corporate managers/owners of nuclear power facilities cannot be trusted to provide, or even possess, accurate, correct, and timely information about nuclear power plant accident status and consequences.
3. Accident consequences do not obey regional, or even national, borders. Fallout from both the Chernobyl and Fukushima accidents extended worldwide. Fallout from Three Mile Island was definitively measured hundreds of miles away.

References

Barringer, F. (1991). Chernobyl: Five years later the danger persists. *New York Times Magazine*.
Brown, K. (2019). *Manual for survival: A Chernobyl guide to the future*. WW Norton and Company.
CTBTO. (2011). *Fukushima-related measurements by the CTBTO*. Retrieved from https://www.ctbto.org/news-and-events/news/fukushima-related-measurements-ctbto
Gorbachev, M. (2006). *Turning point at chernobyl*. Retrieved from https://www.gorby.ru/en/presscenter/publication/show_25057/
Gray, M., & Rosen, I. (1982). *The warning: Accident at three Mile Island*. WW Norton and Co.
Higaki, S. (2023). The discovery of Radiocesium-bearing microparticles directly delivered to a person in Tokyo as a result of the Fukushima Daiichi nuclear disaster. *Health Physics, 125*(5), 325–331. https://doi.org/10.1097/HP.0000000000001719

Higginbotham, A. (2019). *Midnight in Chernobyl: The untold story of the World's greatest nuclear disaster*. Simon & Schuster.

Manzurova, N., & Sullivan, C. L. (2006). *Hard duty: A woman's experience at Chernobyl*.

Medvedev, G. (1991). *The truth about Chernobyl*. Basic Books.

Milnes, A. (2011, April 5). *Jimmy Carter's exposure to nuclear danger*. Retrieved from https://www.cnn.com/2011/OPINION/04/05/milnes.carter.nuclear/index.html?fbclid=IwAR1_Ix_OSRZOfgUVLpJ7kKv5P6sPwIyiQb0MdwDeInPgOBGORnx28TDKU1w

Nadesan, M. H. (2013). *Fukushima and the privatization of risk*. Palgrave Macmillan.

Smith Katagiri, A. M. (1989). *Three Mile Island: The People's testament. On file with the author*. Additionally available on the web at https://www.tmia.com/node/118

Walker, J. S. (2004). *Three Mile Island: A nuclear crisis in historical perspective*. University of California Press.

Wilson, R. T., LaBarge, B. L., Stahl, L. E., Goldenberg, D., Lyamzina, Y., & Talbott, E. O. (2022). What have we learned about health effects more than 40 years after the three Mile Island nuclear accident? A scoping and process review. *Risk, Hazards & Crisis in Public Policy, 14*(2), 129–158. https://doi.org/10.1002/rhc3.12258

Chapter 6
Three Mile Island: An Unresolved Paradox

On May 7, a few weeks after the accident at Three-Mile Island, I was in Washington. I was there to refute some of that propaganda that Ralph Nader, Jane Fonda, and their kind are spewing to the news media in their attempt to frighten people away from nuclear power. I am 71 years old, and I was working 20 hours a day. The strain was too much. The next day, I suffered a heart attack. You might say that I was the only one whose health was affected by that reactor near Harrisburg. No, that would be wrong. It was not the reactor. It was Jane Fonda. Reactors are not dangerous.

> – Dr. Edward Teller, father of the hydrogen bomb, writing in the *Wall Street Journal* (Teller, 1979)

At the time of the TMI accident, I was living ... approximately four miles northwest of TMI. Concerning my experience following the accident at TMI: On Thursday, March 29, 1979, I was working all day with my son in our garage. The garage doors were open. That night when I took a shower, my face, neck, and hands looked like I was at the seashore and got burned real bad. I felt nauseous. My eyes were red and burning. I felt like I was looking through water. Friday morning when I got out of bed, my lips and nose were blistered, and my throat and inside my chest felt like fire. It tasted like burning galvanized steel. My son had similar experiences. He was 22 years old at the time.

> – Affidavit of a resident living near Three Mile Island (Aamodt & Aamodt, 1984).

Becoming involved in the TMI research most certainly changed my life and research options. While emotionally trying I would do it all over again for what I've learned about science, academe, the courts, and the difficult situations of people fighting to overcome a system that exploits rather than serves them.

> – Prof. Steve Wing, University of North Carolina, Chapel Hill; personal communication with the author, 2016.

Aaron Datesman is the primary author of this chapter.

D. Brugge, A. Datesman, *Dirty Secrets of Nuclear Power in an Era of Climate Change*, https://doi.org/10.1007/978-3-031-59595-0_6

The March 28, 1979, meltdown at Three Mile Island (TMI) Unit 2 in Pennsylvania remains the largest nuclear power plant accident, as well the largest industrial disaster, to take place in the Western hemisphere. As such, the topic is necessarily central to any discussion of the role of nuclear power − past, present, and future − in American society. While memory of the accident may be fading, a deep paradox concerning this event remains unresolved. The nature and resolution of this paradox should strongly influence decisions about the future of nuclear power in this country and the world.

The first two quotations opening the chapter illustrate opposing sides of the paradox. The physicist Edward Teller, inventor of the thermonuclear bomb and inspiration for the movie character Dr. Strangelove, asserted that no one could have been harmed by the accident at Three Mile Island.[1] Whatever his expertise, Dr. Teller was not present in Pennsylvania at the end of March, 1979. The observations and opinion of someone who was present at that time and place − whose own health, in short, was an indicator of the severity of the accident − were starkly different. How should the collision between authority and experience be resolved? It is the author's view that the paradox illuminates a profound and worrisome failure of authority. Expert opinion in this instance privileged physical measurement over biological outcomes, and thereby may have failed to recognize and correctly identify injuries (both short- and long-term) due to radiation exposure at Three Mile Island.

By the time an order recommending evacuation for the most vulnerable was issued, two days after the accident began, the damaged facility had already released on the order of 20 million Curies (20 MCi) of the radioactive noble gas xenon-133 (Xe-133) into the environment. As shown in Fig. 6.1, most of the released activity was carried by the wind in a low plume traveling to the northwest. While it is common to read statements to the effect that "very low doses" resulted from the accident (Hatch et al., 1990) the release contained activity equal to that of 20 metric tons (44,000 pounds) of radium, the most radioactive element that occurs naturally. A far smaller amount of iodine-131, around 14 Curies, was also estimated to have been deposited in the ten-mile area surrounding the TMI facility. Some of that radioiodine was discovered in milk collected from local dairy farms.

While the release due to the accident should not be considered small, it is nevertheless correct that the absorbed doses to individuals were not (according to the conventional scientific understanding) alarming. The largest dose

---

[1] It is worth mentioning that the two-page advertisement in the Wall Street Journal in which Dr. Teller made this statement was paid for by Dresser Industries, the manufacturer of the failed valve that was the proximate cause of the accident at Three Mile Island.

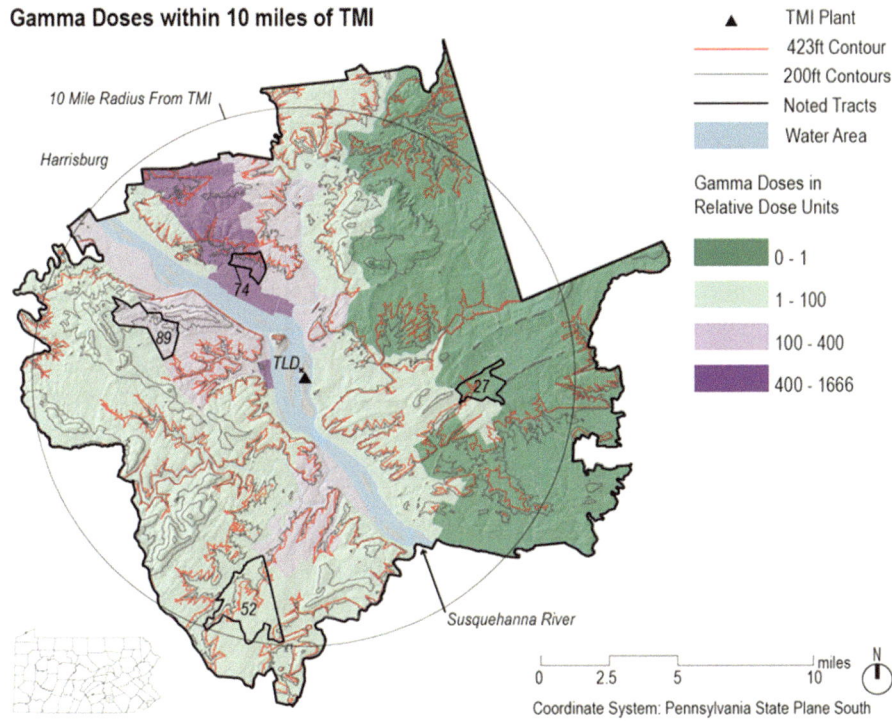

Fig. 6.1 Map of the area surrounding the Three Mile Island nuclear power plant in Middletown, PA. The colors indicate exposure to external gamma radiation due to radioactive xenon released from the auxiliary building vent stack (marked with the filled triangle) during the first thirty-nine hours of the accident. The most intense portion of the plume blew toward the northwest under conditions of steady wind late in the evening on Wednesday, March 28, 1979. This image is drawn from the author's own work (Datesman, 2020)

to any individual due to gamma radiation was estimated to be in the range from 0.7–2 milliSieverts (mSv), which is on the order of the annual dose due to background radiation.[2] Analysis revealed that the total exposure among the two million individuals living within 50 miles amounted to approximately 37 person-Sieverts. The dose would be expected to cause two additional cases of cancer mortality among the affected population. Quite aside from the ethical question centered upon two deaths, at this level of impact no adverse outcomes from the accident should have been observable by means of epidemiology.

Nevertheless, according to a recent review, several independent investigations reported that TMI accident emissions were associated with increased

[2] The comparison to background radiation is significant due to the author's contention that background radiation, while universal, is inadequately understood.

cancer incidence or mortality. The health endpoints[3] included both lung cancer incidence and breast cancer mortality among women (Wilson et al., 2023). If the scientific prediction of no harm were correct, then the epidemiological findings demonstrating harm must somehow be wrong. On the other hand, if the epidemiological findings are correct, then how could the observed medical outcomes result from the low-level exposures that occurred?

There is a pat, but accurate, joke that begins "How do you find five different opinions about a controversial topic?" The answer is, "Ask three epidemiologists." The TMI epidemiological results do not support an unambiguous interpretation, and therefore (like the joke) generate more heat than light. Fortunately, other areas of scientific investigation relating to the accident have the potential to be more illuminating. It is interesting to briefly examine this history.

Dozens of lives have been spent in activism because of the Three Mile Island accident. In some cases, these stories have been captured in records stored in the archives of the library of Dickinson College. The Dickinson archives regrettably do not include the papers of two of the most prolific activists involved with the Three Mile Island issue, Norman and Marjorie Aamodt. An electronic search in the library of the Nuclear Regulatory Commission using "Aamodt" as a prompt, however, returns a request to refine the search terms to limit the results to fewer than one thousand items.

The Aamodts, who were a married couple, possessed interesting backgrounds. Although in 1979 they lived on a cattle farm in Chester County, PA, they were both technically educated. In fact, they met while they were both employed at Bell Laboratories, which at the time was undoubtedly the world's premier scientific research facility. The Aamodts became involved in the TMI issue through a classified advertisement seeking professional expertise in determining the cause of the accident. A lifetime of activism followed, including especially their service to the plaintiff's counsel in the litigation in the United States District Court for the Middle District of Pennsylvania that became known as *TMI Consolidated*. The plaintiffs in the case were more than 2000 persons who believed they had suffered harm due to the meltdown at Three Mile Island.

Speaking at the 1984 Workshop on Three Mile Island Dosimetry, sponsored by the Three Mile Island Public Health Fund, Marjorie Aamodt made the following statement:

---

[3]Among other investigations, researchers at the University of Pittsburgh examined cancer incidence data through 1995, and mortality data through 1998.

....I'm one of the women who did the study on the cancer deaths in the area northwest of the plant. And I would simply like to say that it is not just a matter of how many deaths, but of how much we can learn from the deaths. These people, I believe, were the true dosimeters at the time of the accident. (Beyea, 1985a).

The health survey initiated by Marjorie Aamodt reported cancer clusters in discrete locations northwest of TMI. Between the venue in which Marjorie Aamodt spoke (concerned with dosimetry, that is, a physical measure of exposure) and her statement ("people ... were the true dosimeters"), one finds a clear statement of the central paradox. How do we resolve a conflict between physical measurement and biological outcomes?

The Aamodt survey was an important motivation for a larger, more rigorous epidemiological investigation (including 130,000 persons out to a distance of ten miles) funded by the TMI Public Health Fund. The investigation was undertaken by researchers from Columbia University, under the supervision of the well-regarded epidemiologist Mervyn Susser. Prof. Susser possessed a notable background. He and his wife, Zena Stein, were anti-apartheid activists in their native South Africa, which they left in 1956 due to their political beliefs. Susser went on to become chairman of the division of epidemiology at Columbia in 1966, where he was one of the first epidemiologists to examine the AIDS epidemic when it emerged in New York City during the early 1980's. His profile is not consistent with that of a man willing to act as a toady for a harmful industry.

The results of the Columbia study are often portrayed as providing no evidence for health effects arising from the accident at TMI. This interpretation is not quite correct. In fact, the Columbia researchers found a discernible increase in the incidence of lung cancer among the affected population. Controversy in this matter arises not from the finding itself, but rather, from its interpretation. The Columbia investigators asserted that radioactive emissions from Three Mile Island could not have been causative (Hatch et al., 1990), in part because of the low dose of radiation received.[4] They did not assert that no excess incidence of cancer had been found.

The incorrect impression of the Columbia results has taken root, especially because the plaintiff's attorneys in *TMI Consolidated* engaged their

---

[4] From the paper published by the Columbia team:

... the possibility that emissions from the Three Mile Island nuclear power plant could have contributed to the observed trends, in lung cancer particularly, must be weighed against (1) the lack of effects on the cancers believed to be most radiosensitive and the indeterminate effects on children; (2) the threat of confounding by factors unmeasured or inadequately controlled; (3) inconsistency within our own data between the findings for plant emissions and background gamma radiation; and (4) the low estimates of radiation exposure and the brief interval since exposure occurred.

own expert to re-evaluate the evidence. The expert was Steven Wing, an epidemiologist from the University of North Carolina (UNC) at Chapel Hill and author of the third quotation opening this chapter. Prof. Wing had previously analyzed mortality data among occupationally exposed workers at Oak Ridge National Laboratory. His experience at Oak Ridge left him with a sense of deep skepticism regarding the government/industrial/scientific nexus surrounding nuclear weapons and nuclear energy. Due to his earlier encounter, Wing was initially hesitant to become involved with the TMI litigation. He changed his mind based on the quality of the Aamodt cancer cluster survey, as well as the commitment and reasonableness demonstrated by these two remarkable individuals.

Working with the data collected by the Columbia researchers, Wing and the UNC team reached, for the most part, similar results. Susser wrote precisely this in a letter published in the journal *Environmental Health Perspectives* following the publication of the UNC investigation:

> Our results and those of Wing et al. differ in no important respect. Our conclusions do differ: we saw no convincing evidence that cancer incidence was a consequence of the nuclear accident; they claim there is such evidence. (Susser, 1997)

It is the author's opinion that this statement is a valid summary of the Columbia-UNC controversy. Unlike the Columbia researchers, the UNC team was willing to assert that TMI accident emissions could have caused the increased incidence of lung cancer. Their willingness to make this controversial claim seems mostly to have been due to three factors. The UNC team a) made no assumption that the doses were "low level", b) analyzed the data with more granularity, and therefore were more strongly convinced by the dose response for lung cancer, and c) gave more weight to anecdotal evidence of radiation exposure.

In short, two well-credentialed teams of epidemiologists using the same data reached similar results, but nevertheless interpreted those results in diametrically opposite ways. The reader might be excused for concluding that epidemiology is not a rigorous scientific discipline. However, in the author's opinion, there is a more subtle lesson. The conclusions to be drawn from epidemiology can only be as robust as the underlying understanding of the physical and biological mechanisms connecting exposure to harm.

The insight Marjorie Aamodt expressed at the workshop in 1984 is therefore valuable. If epidemiological results cannot be interpreted reliably due to a possibly deficient physical understanding, a biological dosimeter – that is, a yardstick by which to directly measure the impact of exposure on a living organism – might shed light instead. Such a yardstick does exist; in fact, its nature and use had already been described by two scientists from Oak Ridge National Laboratory in 1962. The relevant scientific field is known as cytogenetics, that is, having to do with the structure and function of human chromosomes. Chromosome "aberrations" due to DNA misrepair are indicators of the severity of exposure to ionizing radiation.

Marjorie Aamodt was not alone in her assessment. The Advisory Panel on Health Research Studies established by the Pennsylvania Department of Health (DOH) in 1979 initially called for a program in cytogenetic dosimetry as one of several recommended investigations. Citing "uncertainty surrounding causes of DNA strand breakage," however, the panel later reconsidered its position (Wilson et al., 2023). It does not appear that the cytogenetic investigation recommended by the PA DOH was ever conducted.

Nevertheless, a human cytogenetic (that is, biological) investigation was eventually performed. In 1994–95, a Russian scientist named Vladimir Shevchenko twice visited Central Pennsylvania while engaged as an expert witness for the plaintiffs in *TMI Consolidated*. His participation was prompted by Norman Aamodt, whom he had met at a scientific conference in Geneva, Switzerland in 1994. Dr. Shevchenko had trained in ecology in the Soviet Union. His specific area of expertise involved assessing damage to forest ecosystems due to radioactive contamination. His work had taken him[5] to the sites of multiple radiological disasters across the Soviet Union, including western Siberia near the Semipalatinsk test site, the area near the Mayak site where plutonium was manufactured, and to Chernobyl in Ukraine, where a reactor lacking a containment structure exploded in 1986. Norman Aamodt had the privilege of spending many days shuttling Dr. Shevchenko around the area around Harrisburg, PA, looking at trees.

Cores were taken from more than eighty trees in Pennsylvania, which were sent back to Russia for analysis. It was Shevchenko's professional scientific opinion, conveyed in an official report filed with the District Court for the Middle District of Pennsylvania, that damage to trees indicated exposures in the range from 2000–10,000 mSv in locations northwest and west of TMI where the plume of xenon had been most intense. Shevchenko was not alone in his assessment. The American scientist Dr. James Gunckel, who was a world authority on modifications of plant growth and development induced by ionizing radiation, had expressed a concordant opinion about a decade earlier. Dr. Gunckel had examined deformed plants such as those shown in Fig. 6.2. He wrote the following in 1984:

> I have carefully examined a few specimens of common plants collected shortly after the accident at TMI and compared them with specimens collected more recently. The current abnormalities are probably carried forward by induced chromosome aberrations ... it would

---

[5] Regrettably, Dr. Shevchenko passed away of stomach cancer at a rather young age, circa 2005. When the author spoke to him in 2018, Mr. Aamodt still recalled the time he had spent with Dr. Shevchenko years before with obvious fondness and respect.

Fig. 6.2 An example of a mutated plant observed by local resident Mary Osborne. The image here was included in the records of the 1985 Workshop on Three Mile Island Dosimetry. From (Beyea, 1985b)

have been possible for the types of plant abnormalities observed to have been induced by radioactive fallout on March 29, 1979. (Aamodt & Aamodt, 1984)

The results with plants are significant because they contradict the conventional explanation offered by the authorities. A psychological explanation for some of the adverse health impacts experienced by human beings, as the authorities assert, is not implausible. Both the accident, and the subsequent evacuation, were very traumatic events in the lives of the individuals affected. The psychological explanation, however, is incompatible with the observation of injury to trees.

Dr. Shevchenko did not confine his on-the-ground investigation to plant life. He also interviewed people living nearby. In the same locations where the structure, growth, and health of trees indicated exposure to ionizing radiation, he wrote, "the residents in these areas felt at the time of the accident unusual events about their health," including.

> ...skin redness and rashes, nausea, inflammation of the eyes, metallic taste, inflammation of respiratory ways, diarrhea, anal bleeding, hair loss, interruption of the menstrual cycle, pain in the joints, and others. (Shevchenko, 1995)

The symptoms in humans, Dr. Shevchenko wrote, were consistent with radiation sickness resulting from an exposure in the range of 1000 mSv. The assessed doses to trees are higher than the doses to human beings both because the xenon plume was elevated, and also because the living portion of a tree (the bark and leaves) is external, and therefore unshielded.

Dr. Shevchenko additionally coordinated a sizable investigation that drew upon experts from disparate fields (including botany and ecology, physical dosimetry, immunology, and cytogenetics) from within the Russian scientific establishment. The final report of the investigation he oversaw is troubling, and almost completely unknown. This was the finding as expressed by Shevchenko:

> In the cytogenetic report the data on the level of dicentrics[6] in residents living around TMI is compared to the results of cytogenetic investigation of populations exposed to irradiation .... approximately the same frequencies of dicentrics were found out in residents of the areas around TMI and the residents of a number of regions in Russia most suffered (sic) from the action of ionizing radiation. (Shevchenko, 1995)

Summarizing the findings concisely, this is what Dr. Shevchenko found: the level of biological damage among the persons examined in Central Pennsylvania was comparable to that discovered among members of the Altai population in Western Siberia who were severely exposed to fallout from an atomic bomb. The finding is consistent with observations of damaged and deformed trees and plants, anecdotal information gained from interviews, and evidence of immunological deficiencies (among other insights), but at the same time seemingly inconsistent with the observation that the TMI exposures were about the same as the annual dose due to background radiation.

The comparison returns the discussion to its central theme, the contradiction between physical measurement and biological outcomes. In the author's opinion, it is likely that the following two observations are both simultaneously correct:

1. The absorbed dose to any individual around TMI was small (less than about 2 mGy). The conclusion is anchored to physical measurements, most significantly those made using electronic devices known as thermoluminescent dosimeters (TLDs).
2. The biological impact to the most-exposed individuals was severe (in the range of 600–1000 mSv). The conclusion is supported by biological outcomes: anecdotal evidence consistent with radiation exposure, the outcomes of the investigations performed by Russian scientists, and (although not conclusive) epidemiology.

It is of course legitimate to view the results of the investigations coordinated by Shevchenko – which took place in the context of a legal proceeding, and

---

[6] A dicentric is a variety of chromosome aberration.

which were never published in the peer-reviewed literature – with a reasonable degree of skepticism. Might they be mistaken, incorrectly interpreted,
or manipulated in some way? The most sensible response to the concern is
to replicate the investigation. Because some chromosome aberrations are
stable over time – that is, they exist for the entire lifetime of the exposed
person – this possibility remains.

Recognizing the possibility, the authors of this book (along with other
collaborators) have undertaken just such an investigation. A karyogram from
our 3MILER RUN (Three Mile Island Low level Exposure to Radioxenon:
a Re-assessment Using New cytogenomics) investigation is shown in
Fig. 6.3. The results of our preliminary investigation should be available by
late 2024.

The paradox embedded in the story of the Three Mile Island accident, it
has been argued, represents the dichotomy between physical measurement
and biological outcomes. Proponents of nuclear power technology are, for
the most part, anchored in the community of engineering and the physical

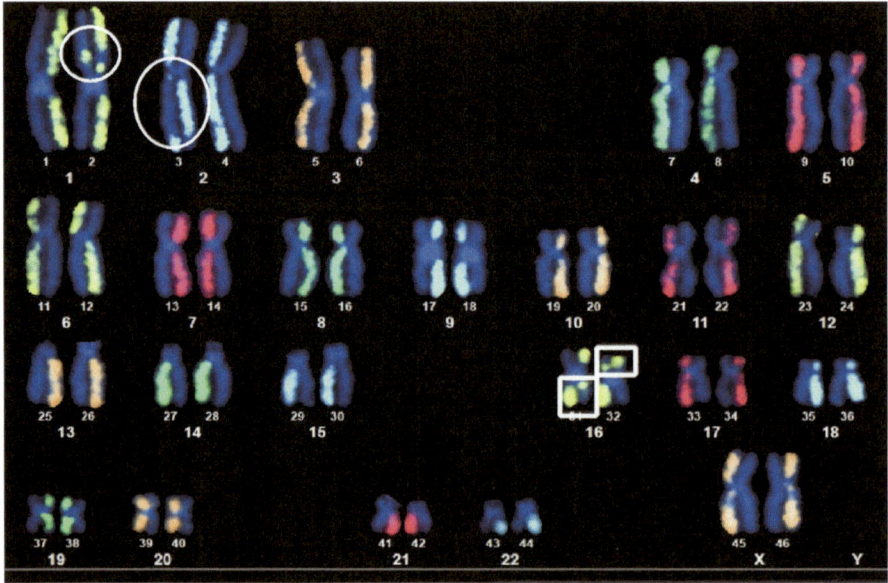

Fig. 6.3 A karyogram (an image of the complete set of chromosomes from one person) from the
3MILER RUN investigation. The study participant to whom this karyogram belongs lived within
ten miles of TMI in 1979. Because the investigation remains blinded as this book goes to press, the
authors are not aware whether this individual was exposed to radioactive xenon. The total number
of chromosome aberrations (marked with circles in the image) appearing in hundreds of such
karyograms belonging to one individual may provide indication of biological harm due to radiation
exposure. The technique is similar to that used for the 1994 cytogenetic analysis, although techno-
logically more refined

sciences. Because the culture is technophilic, and often quite callous regarding questions of risk, and because the weight of power and wealth are on the side of government and industry, the side of the dichotomy anchored to the physical understanding has dominated official perspectives. Overall, the power imbalance acts to exclude the observations of affected persons regarding their own health, in favor of a mere physical theory.

Regarding the future of nuclear power, the lesson of the Three Mile Island accident in the present day ought to be that questions of harm from low-level exposure are not yet settled. If the degree of harm resulting from "low-level" exposures is in fact not negligible, then additional engineering controls will be necessary to mitigate these previously unrecognized or discounted risks. The necessary changes would increase the already uncompetitive costs of nuclear power technology.

Summary Points

1. Injury caused by exposure to ionizing radiation may be assessed using methods anchored either in physical or biological understanding. There is significant biological evidence of severe harm due to the low level TMI exposure.
2. A cytogenetic investigation named 3MILER RUN, conducted by the authors, is underway at the time of publication.
3. If biological indications that the TMI exposures were harmful were in fact correct, the additional engineering controls necessary to construct new nuclear power stations to an acceptable level of safety will almost certainly be prohibitively expensive.

References

Aamodt, M., & Aamodt, N. (1984). Petitioners v. U.S. Nuclear Regulatory Commission. Aamodt motions for investigation of licensee's reports of radioactive releases during the initial days of the TMI-2 accident and postponement of restart decision pending resolution of this investigation. Docket Number 50-289. Administrative Court, Washington, DC, 21 June.

Beyea, J. (1985a). *Proceedings of the Workshop on Three Mile Island Dosimetry* (Vol. 1, p. 124). Three Mile Island Public Health Fund and the Academy of Natural Sciences of Philadelphia.

Beyea, J. (1985b). *Proceedings of the Workshop on Three Mile Island Dosimetry* (Vol. 2, p. B129). Three Mile Island Public Health Fund and the Academy of Natural Sciences of Philadelphia.

Datesman, A. M. (2020). Radiobiological shot noise explains Three Mile Island biodosimetry indicating nearly 1,000 mSv exposures. *Scientific Reports, 10*(1), 10933. https://doi.org/10.1038/s41598-020-67826-5

Hatch, M. C., Beyea, J., Nieves, J. W., & Susser, M. (1990). Cancer near the three-mile Island nuclear plant: radiation emissions. *American Journal of Epidemiology, 132*(3), 397–412. https://doi.org/10.1093/oxfordjournals.aje.a115673

Shevchenko V. (1995). *The final report of Prof. Vladimir A. Shevchenko, Ph.D., Dr. Sc., concerning the dose to any individual from the TMI Unit 2 accident.* Copy on file with the author.

Susser, M. (1997). Consequences of the 1979 three mile Island accident continued: Further comment. *Environmental Health Perspectives, 105*(6), 566–567. https://doi.org/10.2307/3433589

Teller, E. (1979). I was the only victim of the Three Mile Island. *The Wall Street Journal, 31*, 1979.

Wilson, R. T., LaBarge, B. L., Stahl, L. E., Goldenberg, D., Lyamzina, Y., & Talbott, E. O. (2023). What have we learned about health effects more than 40 years after the Three Mile Island nuclear accident? A scoping and process review. *Risk, Hazards & Crisis in Public Policy, 14*(2), 129–158. https://doi.org/10.1002/rhc3.12258

Chapter 7
Protracted Exposures May Be Misunderstood

The most important source for evaluating the dangers of radiation to large population groups was the study of survivors of the atomic bombings of Hiroshima and Nagasaki. It supplied the best available epidemiological data on the effects of radiation on humans, and scientific knowledge about radiation hazards drew in significant measure from the work of the Atomic Bomb Casualty Commission.... (Walker, 2000)
> – J. Samuel Walker, historian of the U.S. Nuclear Regulatory Commission.

The A-bomb studies have set standards that are patently false. (Greene, 2003)
> – Dr. Alice Stewart, pioneering British epidemiologist, pictured in Fig. 7.1.

The Postal Analogy

Imagine a vast postal sorting warehouse. Under one roof, extending to the horizon, there sits an enormous collection of conveyor belts, each filled with corrugated postal bins. Perhaps each conveyor belt corresponds to a different location, while every postal bin is directed to a certain truck leaving at a certain time. The bins move by continuously, independently, on each belt, and very quickly. The scale of the entire operation is immense. For the purposes of illustration, consider just a small portion of the entire warehouse,

---

The following chapter covers conceptual material with a mathematical basis, which may not be of universal interest. If the reader grasps the fundamental contention that protracted low-level exposures to ionizing radiation may be inadequately understood, there will be no harm to skipping to the summary points at the end.
Aaron Datesman is the primary author of this chapter.

© The Author(s) 2024
D. Brugge, A. Datesman, *Dirty Secrets of Nuclear Power in an Era of Climate Change*, https://doi.org/10.1007/978-3-031-59595-0_7

Fig. 7.1 Dr. Alice Stewart, the epidemiologist who demonstrated that x-rays in utero cause leukemia in young children. The author fervently recommends the brilliant biography of Dr. Stewart written by Gayle Greene, <u>The Woman Who Knew Too Much</u> (Greene, 2003)

shown in Fig. 7.2. Within the field of view, ten conveyor belts are arranged. On each conveyor belt, at one instant in time, there sit ten bins. A total of one hundred bins are visible.

A mail sorter has 3000 pieces of mail to distribute among these 100 bins. The mail is all junk mail, it doesn't matter where it goes, and the sorter distributes the individual pieces of mail at random. How many pieces of mail wind up in each bin? It is obvious that, as an average matter, each bin will contain 30 letters. Because the process is random, however, some bins will contain more than the average number, and some less. A valid distribution[1] is shown in the leftmost grid of Fig. 7.2. Despite the random distribution between individual bins, on average the ten postal bins in one row (on one conveyor belt, corresponding to one location) together contain about 300 pieces of mail.

It is interesting next to extrapolate downward. What if there are only thirty letters to distribute among these one hundred bins? Or three? Because a letter may not be cut into pieces, many - or most - bins are now empty. (It may be junk mail, but it may not be delivered pre-shredded!) These situations are illustrated in the middle and rightmost grids of Fig. 7.2.

The alert reader may ask, what about an extrapolation below one letter? Is there a lower limit? Because the number of conveyor belts is very large but (in theory) not fixed, the answer is that no lower limit exists. To extrapolate further downward, it is necessary simply to expand the field of vision to

---

[1] The statement means specifically that the distribution among bins obeys "Poisson statistics".

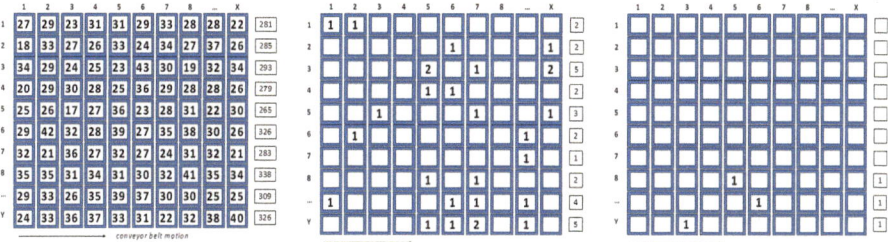

Fig. 7.2 The postal sorting warehouse contains an ensemble of Y conveyor belts, each holding X postal bins. Ensembles representing high (left), intermediate, and low (right) dose rate exposures are illustrated. The number of pieces of mail distributed are 3000, 30, and 3, respectively. From left to right, the physical picture changes from describing the number of events contained in each bin, to describing how often a bin contains a single event. The rightmost column states the sum of the number of pieces of mail across the row. Because the postal bins on one conveyor belt represent divisions of time, the illustrated ensembles transition from "temporally homogeneous" on the left to "temporally inhomogeneous" on the right

include more conveyor belts.[2] In statistical physics, the collection of postal bins on an infinite number of conveyor belts that has been conjectured is called an "ensemble". It is never necessary to cut apart a piece of mail to represent any desired average number of pieces of mail on one conveyor belt (300, 3, 0.3, 0.0003, etc.) if the ensemble is infinite.

Moving from left to right in Fig. 7.2, one observes that a meaningful description of the three different distributions transitions from describing "how many" pieces of mail are contained in each bin, to "how often" each bin contains a single event. In technical terms, the distinction is that between the amplitude of events, and the frequency with which they occur. Human beings, especially due to the tendency to discount the occurrence of improbable events, reason most comfortably in the realm of "how many". Nature, however, incorporates both viewpoints simultaneously. The correct mathematical construct is known as the "ensemble average". Human common sense unfortunately misleads in this situation.

The numerical distinction between the average and the ensemble average, illustrated with reference to the examples shown in Fig. 7.2, is presented in Table 7.1. When the postal bins are very full, there is little difference between the average value, and the ensemble average value. There is nevertheless a very important difference between them: while the average value extrapolates linearly downward, the ensemble average value does not.[3] (Two

---

[2] To take a concrete example: if the average number of pieces of mail per bin is 0.0001, a valid random distribution would arise from sorting 100 pieces of mail among one million bins, arranged on ten thousand conveyor belts.

[3] Taking n as the average number of events in one member of the ensemble (representing the absorbed dose), the total number of events in the ensemble is $N = nY$. The ensemble average

Table 7.1 Illustration of the distinction between average and ensemble average values using the ensembles presented in Fig. 7.2

Number of letters	Average	Range	Ensemble average
3000	300	265–338	305
30	3	1–5	6.2
3	0.3	0–1	1.8
	0.0003	0–1	0.06

"Average" refers to the average number of pieces of mail in one row of the grid. The average is an analogous quantity to the absorbed dose. "Range" refers to the lowest and highest number of mail pieces delivered in any one row. Note that the range cannot extrapolate downward below its discrete unit of one letter. The bottom row refers to a separate ensemble with a very sparse distribution, which would have to be too large to illustrate usefully

quantities are linearly related if separated by a constant factor; for instance, the "Number of Letters" and "Average" columns in Table 7.1.) Instead, the ensemble average exceeds the average value by a large factor that increases as the grid becomes filled more sparsely.

The reader may have a difficult time accommodating to the concept of the ensemble average, which after all invokes some complex ideas. The fundamental issue, however, is concrete: the Post Office is not allowed to shred a piece of mail with the goal of placing the same weight of mail in every postal bin. The distribution of letters is not uniform. ("Inhomogeneous" is a useful description.) The random process of distributing mail bumps up against a discrete lower limit, as may be seen in the bottom two rows of Table 7.1. The ensemble average represents Nature's compromise between the decreasing average value (for instance, 0.0003 letters, a felony act according to the U.S. Postal Code), and the lower limit imposed by the discrete nature of a letter (1 piece of mail, indivisible).

The foregoing discussion has been an attempt to construct by analogy a simple description of a complex phenomenon. That phenomenon is biological injury due to exposure to ionizing radiation. Each piece of mail represents a single ionization event - that is, the liberation of a single energetic electron in living tissue. (The liberation of an electron is the reason the phenomenon is referred to as "ionizing" radiation.) Every ionization event has the possibility to be followed by biological injury. The distribution of events

---

number   of   events $\bar{n}$ for   the   sparse   case   is

$\bar{n} = X{\cdot}RMS = X{\cdot}\sqrt{\sum_{i,j} N_{ij}^2 / XY} = X{\cdot}\sqrt{N / XY} = X{\cdot}\sqrt{nY / XY} = \sqrt{nX}$, where RMS indicates the

Root Mean Square average number of events in one bin. The complete expression encompassing both "viewpoints" is $\bar{n} = \sqrt{nX + n^2}$, which has a linear form if n is large but a non-linear form if n is small.

is random, both in time and in space. The assignment of events to bins represents an understanding that events occur both in a certain location (on one conveyor belt, rather than another) and at a certain time (in a certain postal bin). The rate at which the postal bins on a conveyor belt whisk by represents the rate of the chemical reaction responsible for biological injury. The total number of bins on one conveyor belt represents the duration of exposure.

Adding up across the rows the packets of energy associated with individual ionization events to arrive at the total "dose", commonsensically, indicates the severity of exposure. Since both the amplitude and the frequency of events must be accounted for, however, Nature judges the situation with more subtlety, using the ensemble average. It follows that a protracted, low-level exposure to ionizing radiation may be far more damaging to health than an acute exposure depositing the same absorbed dose in tissue.

> The Gray (Gy) is a physical measure of the *absorbed* dose, indicating the quantity of energy dissipated by ionizing radiation in a volume of tissue. The unit for the *equivalent* or *reference* dose, which is a biological measure of risk, is the Sievert (Sv). It is believed that an acute dose of 1 Sv increases the risk of developing a fatal cancer by 5.5%. Referring to the postal bin analogy, the author contends that the average value corresponds to the *absorbed* dose in Gray, while the ensemble average value corresponds to the *equivalent* dose in Sieverts. The hypothesis goes by the name of "shot noise in radiobiological systems".

Challenging the Linear Model

On the question of radiation protection, and of the impacts to human health of low-level exposures to ionizing radiation, expert opinion endorses what is known as the "Linear No Threshold", or LNT, model. The National Academy of Sciences most recently supported this outlook in its 2006 report, titled "Health Risks from Exposure to Low Levels of Ionizing Radiation" (BEIR VII). The LNT model asserts two essential findings. First, according to the LNT model there is no such thing as a "safe" dose. Any single interaction may be biologically damaging, and there is furthermore no evidence supporting any threshold of exposure below which damage cannot occur. The second assertion made by the LNT model is linearity: it is believed acceptable to extrapolate impacts linearly downward from high doses to low doses. The extrapolation is necessary because it is difficult to directly evaluate health impacts resulting from exposures of less than about 0.1 Gy.

The BEIR VII report justifies the linear extrapolation on the following basis:

[A]ny single track of ionizing radiation has the potential to cause cellular damage. However, if only one ionizing particle passes through a cell's DNA, the chances of damage to the cell's DNA are proportionately lower than if there are 10, 100, or 1000 such ionizing particles passing through it. There is no reason to expect a greater effect at lower doses from the physical interaction of the radiation with the cell's DNA. (National Research Council, 2006)

The postal bin analogy from the previous section has prepared the reader to consider the argument offered by the BEIR VII committee. While intuitively sensible and therefore very appealing, the argument for linearity in fact is valid only under very special circumstances. The difficulty arises because there exist proper scales of volume and time describing physical/chemical/biological action. In short, what size are the postal bins? Yes, ten, one hundred, or even one thousand or more events that may cause injury are posited to occur, but within what interval of time? Within what volume of tissue? Nature has very specific answers to these questions, which however expert opinion does not identify.[4]

Because the authorities have never adequately conceptualized the implicit question about natural scales of time and volume, in the author's view the authorities have also failed to recognize the necessary role of the ensemble calculation. The language chosen by the BEIR VII committee, which adopts the frame of "how many?" events while failing to address "how often?" events occur, supports this interpretation. While the LNT model permits outcomes to extrapolate downward, it encompasses no mechanism to extrapolate exposures downward below the level of one track per cell. The difficulty is a serious one. However, the resolution to this contradiction can be addressed as it was in the postal analogy.

Nature utilizes the ensemble average value, which reflects both the amplitude and the frequency of events. Only the amplitude of events, represented by the average value, extrapolates linearly downward. When every bin is full (a temporally homogeneous exposure), it happens that the average value about coincides with the ensemble average value. For this reason, the LNT model in this situation agrees with the ensemble average value. In sparse (temporally inhomogeneous) ensembles when many bins are empty, however, the LNT model and the ensemble average value diverge. The LNT model is therefore incorrect for protracted, low dose rate exposures.

---

[4]The chemical species most responsible for radiation injury is the hydroxyl radical, OH. The hydroxyl radical is highly reactive, with a very short lifetime in tissue of only 1 nanosecond (0.000000 001 s). The interaction volume of approximately 0.3 milliliters corresponds to photoelectric absorption of the 1.460 MeV gamma ray produced by potassium-40, a radioactive contaminant universally present in living tissue.

To recap: in a single interaction volume during a single interval of time, either an ionization event occurs (possibly several), or it does not. Each event is discrete, occurring in a single bin representing a single interaction volume during a very brief interval of time. The mathematical representation of the process never treats an event as though it can be divided. The scientific heritage of this idea is more than a century old, having first been developed to describe the failure of vacuum tubes to operate optimally at low levels of amplification. The theory as applied to radiation injury is known as "shot noise in radiobiological systems" (Datesman, 2016).

From the localized perspective of affected tissue, one should conclude from this description that there is no such thing as a "low dose" or a "low-level exposure". Every individual ionization event creates, in a limited volume, a high dose rate exposure of very short duration. Because an overall exposure of finite duration is built up from a series of discrete events as shown in Fig. 7.3, all exposures should therefore be viewed as high dose rate exposures. Exposures characterized as "low dose" are built up from individual high dose rate ionization events, spread out over increasing intervals of time. As a consequence, the LNT model vastly understates the chemical, biological, and medical impact of dilute, protracted exposures to ionizing radiation. The threshold dose rate at which the LNT model begins to fail—because it attempts to extrapolate linearly downward beyond the level of a unit event per bin—lies in the regime below about 100 Gy/hr.

No argument has been made that lower-dose/dose rate exposures are more damaging, although it follows from the shot noise hypothesis that protracted exposures are more damaging per unit of absorbed dose. The speed of DNA repair (it takes about 2 h to repair a double strand break) also has complex consequences for radiation injury in the case of prolonged exposure.

Fig. 7.3 A low dose exposure consists of a series of very brief, high dose rate exposures, schematically illustrated as orange "pulses", spread out in time. In electrical engineering, a waveform of this kind is known as a "pulse train"

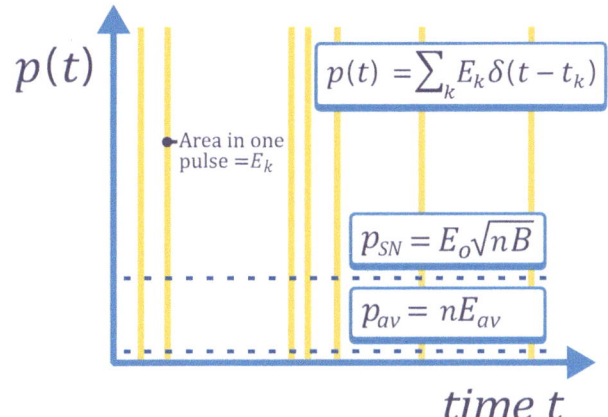

$$p(t)$$

$$p(t) = \sum_k E_k \delta(t - t_k)$$

Area in one pulse $= E_k$

$$p_{SN} = E_0 \sqrt{nB}$$

$$p_{av} = nE_{av}$$

time t

If the shot noise phenomenon is indeed firmly established, so that the LNT model contradicts physical law, how is it that the important knowledge described in this section has been overlooked for many decades? The interesting answer is that it has not. The excerpt below is taken from a prominent lecture delivered by Harald Rossi, the primary inventor of the field of microdosimetry:

> Nearly all physical quantities are nonstochastic, although in many instances the discreteness of matter and of radiation causes statistical fluctuations. However, in most cases these are small enough to be ignored, and no attempt is made to consider the underlying stochastic quantity or to give it a special name. (Rossi, 1986)

Rossi continues to provide a very clear description of the nature of shot noise in electrical circuits, which it is not necessary to reproduce here. The important point is that, although Rossi possessed a clear understanding of the existence, nature, and relevance of shot noise, his conclusion was incorrect: the statistical fluctuations (meaning whether adjacent bins are full or empty) are by no means negligible. The consequences of this error are significant and concerning.

Conventionally, the absorbed dose in Grays and the equivalent dose in Sieverts are related by a "radiation weighting factor," derived using microdosimetry. The weighting factors permit radiations of different "quality" – meaning x- and gamma rays, beta particles, alpha particles, and neutrons – to be compared on the basis of biological injury. It is the author's opinion that the hypothesis of shot noise in radiobiological systems may invalidate the concept of the radiation weighting factor.

Experimental Evidence for the Hypothesis of Shot Noise in Radiobiological Systems

In 1972, Dr. Abram Petkau of the Whiteshell Nuclear Research Establishment in Canada published the results of an intriguing experiment in the mainstream scientific journal *Health Physics*. Dr. Petkau was the head of the Medical Biophysics branch at his institution, affiliated with the Canadian atomic energy establishment. While investigating topics related to radiation chemistry and biological injury, Petkau had previously devised a method to create artificial biological membranes in a small apparatus, in which the membrane spanned an aperture separating two compartments filled with water. With this arrangement it was possible to irradiate the water with x-rays, while simultaneously observing the membrane under a microscope.

It was found that the membranes reliably ruptured (an observable biological outcome) after an absorbed x-ray dose of 35 Gray.

The rupture dose was not found to be replicated when external x-ray irradiation of the membrane was replaced with irradiation from a beta-emitting radionuclide, dissolved in the water contained within the apparatus. It was found instead that membrane rupture occurred at far smaller absorbed doses, increasing with increasing dose rate (Petkau, 1972). The result is intriguing because the Petkau experiment is a direct interrogation of the concept of different qualities of ionizing radiation. Far from confirming the accepted belief that x-rays and beta particles are of identical quality, irrespective of dose rate, the Petkau experiment indicated that beta particle irradiation was as much as 3000 times more effective for membrane rupture.

In the context of the discussion in this chapter, it is noteworthy that Petkau's experiment investigated low dose rates, in all cases less than 0.6 Gy/h. This value lies far below the threshold of approximately 100 Gy/hr. at which the author asserts the linear extrapolation begins to break down. For this reason, the Petkau experiment directly interrogates not only the concept of the radiation quality (which it appears to invalidate), but also the hypothesis of shot noise in radiobiological systems. As shown in Fig. 7.4,

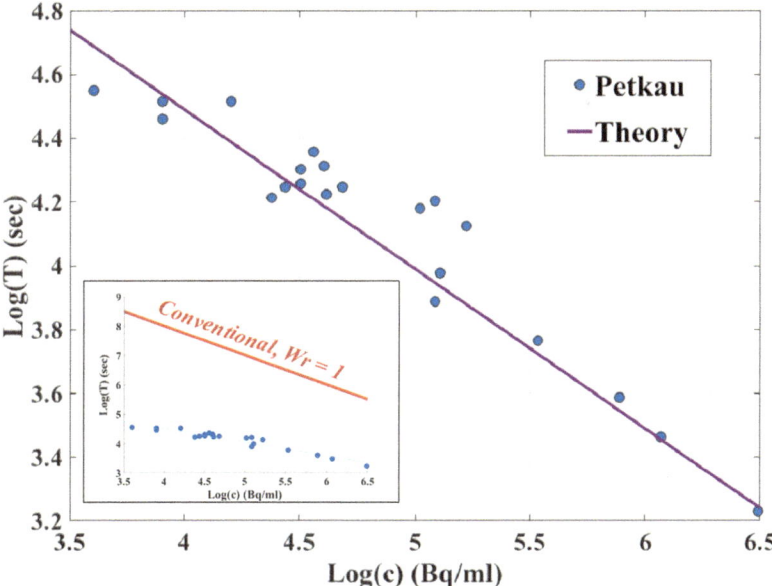

Fig. 7.4 The Petkau Effect explained by the hypothesis of shot noise in radiobiological systems. The membrane rupture time is plotted on the vertical axis, while the concentration of the beta-emitting radionuclide sodium-22 is plotted on the horizontal axis. From the author's own work (Datesman, 2019)

the Petkau result agrees with the prediction of the shot noise hypothesis reasonably well. The finding that the membrane rupture result is consistent with the hypothesis of shot noise in radiobiological systems was published by the author in an article that appeared, also in *Health Physics*, in 2019.

Background Radiation May Not Be Properly Understood

Environmental and medical exposures comprise the so-called "background radiation". Environmental exposures especially are generally protracted in character. At the current time, in the United States the average annual expo-sure to natural (non-medical) sources of ionizing radiation totals about 3 mSv/yr.[5] The contribution to background due to medical radiation (mostly CT scans and x-rays) is of similar magnitude. There are four categories of exposure to non-medical background radiation: inhalation, ingestion of food and water, terrestrial radiation, and cosmic radiation. The inhalation route (principally radon) accounts for 2.3 mSv/yr., while the other contributions all lie in the range of 0.2–0.3 mSv/yr. The stated values are merely averages, as there are large local variations due to geology, building construction, alti-tude, and personal factors.

A particularly interesting contributor to the background dose is potas-sium-40 (K-40), listed in the first row of Table 7.2. Living tissue and blood are universally contaminated with this long-lived radioisotope, which is an

Table 7.2 Comparison of low dose exposures using the hypothesis of shot noise in radiobiological systems

Source	Dose (mGy)	Duration	Avg. Rate (Gy/hr)	Ens. Avg. Rate (Sv/hr)	Eq. Dose (mSv)
Potassium-40	0.15	1 year	0.000000 02	0.0015	13,000
Chest x-ray	0.15	0.1 s	5	23	0.64
LSS A-bomb	39.6	10 s	14	40	110

Acute exposures obey the LNT model, but protracted exposures diverge from the model beginning about at the threshold dose rate. "LSS" refers to exposures to delayed gamma radiation from the Life Span Study of the Japanese survivors of the atomic bombings. The Average Rate corresponds to the absorbed dose rate, in Grays/hour, while the Ensemble Average Rate corresponds to the equivalent dose rate indicating the degree of biological injury, measured in Sieverts/hour. The Equivalent Dose is the product of the exposure Duration and the Ensemble Average Rate. Values of 100 Gy/h. are assumed for the threshold dose rate except in the case of K-40, where it is possible to calculate a value of 134 Gy/h. precisely

---

[5] In this discussion the use of the unit Sievert (Sv) is the conventional one, rather than the novel definition reflecting the ensemble average of the absorbed dose. It is necessary to make this accom-modation because radon is an alpha-emitting radionuclide, with a weighting factor of 20.

energetic mixed emitter (that is, it emits both gamma rays and beta particles). About 4000 decays of K-40 occur in a human body of 60 kg every
second. The absorbed dose due to this contaminant comes to about 0.15 mGy/
year, corresponding to an absorbed dose rate of 2×10^{-8} (or twenty one-
billionths) of a Gray per hour. Invoking the postal warehouse analogy, this
low environmental dose rate corresponds approximately to one postal bin
out of every 56 billion containing a single event. The ensemble in this case
is indeed filled very sparsely. The ensemble average dose rate, reflecting the
degree of biological injury listed in the fifth column of Table 7.2, exceeds
the absorbed or average dose rate by a factor of about 100,000 times.

Medical exposures for diagnostic purposes are generally of reasonably
short duration. A notional chest x-ray delivering a dose of 0.15 mGy with a
duration of 0.1 s is considered in the second row of Table 7.2. In this case,
again because the dose rate is lower than the threshold value, the ensemble
average dose rate (representing the biological impact of the exposure) is also
larger than the absorbed dose rate, although only by a factor of about four.

As explained by the official historian of the Nuclear Regulatory
Commission in the quotation opening this chapter, the Life Span Study
(LSS) of survivors of the atomic bombings of Hiroshima and Nagasaki is the
principal foundation upon which rests our knowledge regarding the harms
of low dose exposure to ionizing radiation. Findings from the LSS essentially anchor the scale society uses to understand the hazards of exposure to
ionizing radiation. The third line of Table 7.2 displays the average absorbed
dose represented in the LSS due to "delayed gamma radiation" for individuals 2000 meters from the hypocenter at Hiroshima.

Table 7.2 reveals an interesting finding.[6] Between the chest x-ray and the
LSS exposures, the absorbed doses differ by a factor of 264. The equivalent
(biological) doses, meanwhile, differ by a similar factor of 172. The similar-
ity indicates that, although the exposure durations differ by a large factor,
the scale established by the LSS describes diagnostic medical radiation
exposures reasonably well.

One should not mistake coincidence for wisdom, however. If the hypothesis of shot noise in radiobiological systems is correct, the framework based
upon the LSS data is fundamentally unsuitable as regards environmental
exposures. As shown in the first row of Table 7.2, the equivalent dose due to
potassium-40 amounts to about 13,000 mSv per annum. For comparison, an
acute dose (that is, occurring in a short duration) in the range of 5000 mSv
will be fatal for a majority of those exposed within 30 days.

---

[6] The technical term describing this situation is "Correct, But For the Wrong Reasons," which
despite the utility of the expression does not make a good acronym.

What could this surprising result mean? Contamination with potassium-40, because it is both universal and essentially unavoidable, is widely considered to be benign. The judgment may not be totally correct. Instead, a reasonable hypothesis is that the mechanisms of cellular repair have evolved to act at a rate that approximately compensates for the injury to genetic material caused by K-40. (According to Table 7.2, the rate at which that damage accrues is approximately 1.5 mSv/hour.) Biological damage caused by exogenous exposures (whether from an atomic bomb, a medical X-ray, or radon in the basement) occur only on top of the baseline activity of endogenous damage due to potassium-40, and its repair. The conventional view embodied in the LNT model unfortunately elides this complication.

It is the author's opinion that no model of radiation injury describing the biological impact of exposures at environmental dose rates can possibly be valid without a comprehensive temporal description of both damage and repair. The scale built upon the LSS is valid for the simple reason that the exposures against which it is proofed are much shorter in duration (of the order of seconds) than the processes responsible for the repair of double strand breaks (which require hours). The dynamic nature of the repair process is therefore not a necessary component of the description of, for instance, medical x-rays.

When the exposure is continuous and protracted, however, a more thorough analysis incorporating the process of repair must be undertaken. Because the LNT model does not incorporate such an analysis, exposure to "background" levels of ionizing radiation in the environment may be a phenomenon that is incompletely, or incorrectly, understood.

Environmental Releases from Operating Nuclear Power Stations

Emissions of radioactive pollution from nuclear power stations occur both on a continuing basis, as well as in "batch" releases of short duration. For example, for each of the years 1999–2003, operating nuclear power plants worldwide released about 30,000 Curies (Ci) of radioxenon on a continuous basis, and 6000 Ci in batch releases (Kalinowski & Tuma, 2009).

In the United States, utilities are required to report emissions to the Nuclear Regulatory Commission. A portion of a table from one such report, from the Oyster Creek Generating Station in New Jersey, is shown in Fig. 7.5. In each of the first three quarters of 2018, the utility that owns the facility reported that the 636 MW boiling water reactor at Oyster Creek released between about 20–40 Ci of "Fission & Activation Gases", principally the noble gases krypton and xenon. The facility reported no emissions

Period: January 1, 2018 through December 31, 2018   **Unit: Oyster Creek**

A. Fission & Activation Gases	Units	Quarter 1	Quarter 2	Quarter 3	Quarter 4	Est. Total Error %
1. Total Release	Ci	3.89E+01	2.84E+01	2.41E+01	<LLD	24.64%
2. Average Release Rate for Period	µCi/sec	5.00E+00	3.61E+00	3.03E+00	<LLD	
3. Gamma Air Dose	mrad	4.88E-04	3.01E-04	3.95E-04	N/A	
4. Beta Air Dose	mrad	1.88E-04	1.77E-04	1.99E-04	N/A	
5. Percent of ODCM Limit						
- Gamma Air Dose	%	9.76E-03	6.02E-03	7.90E-03	N/A	
- Beta Air Dose	%	1.88E-03	1.77E-03	1.99E-03	N/A	

Fig. 7.5 Emissions from the Oyster Creek Generating Station, as reported to the Nuclear Regulatory Commission by Exelon (Exelon Generation, 2018). The reported doses (in millirads) are of the order of 1–10 µGy (microGray)

for the fourth quarter of 2018, since the reactor was shut down in September of that year.[7]

According to the Code of Federal Regulations, releases of radioactive effluents from operating nuclear power plants are permitted so long as doses to individuals in unrestricted areas do not exceed 0.02 mSv in any hour and 0.5 mSv in a year.[8] Consider therefore a batch release consisting solely of the radioactive noble gas krypton-85, a beta-emitting noble gas. At a highly dilute but constant ambient concentration of 1 µCi per liter, an individual exposed to this release will receive a whole-body gamma dose of approximately 0.02 mSv in 1 h.

While the noble gases are unreactive, they do represent an inhalation hazard. Moreover, the lung epithelial tissue is permeable to krypton, which binds to hemoglobin and is distributed throughout the body via the circulatory system. (The same is also true of xenon, which is important as it relates to the 1979 accident at the Three Mile Island nuclear power station in Pennsylvania.) The activity of krypton-85 in the bloodstream of the individual exposed to this batch release would be 1.7 Bq per milliliter. By way of comparison, the overall K-40 activity in the human body is much less: only 0.067 Bq per milliliter.

For this notional exposure, the absorbed dose rate to the blood would be 245 nGy/hr. If the hypothesis of shot noise in radiobiological systems is correct, however, this permitted exposure to Kr-85 would have a biological impact greater than 5 mSv. A statement of risk places the result into context. Extrapolating the calculated individual dose up to the population level, and

---

[7] In September 2019, one year later, the developer of the state of New Jersey's first offshore wind farm secured capacity interconnection rights at the former Oyster Creek site. It may be in the future that the infrastructure built to connect a nuclear power plant to the electrical grid will instead supply electrical energy generated by wind power.

[8] The relevant law is 10 CFR 20, Standards for Protection Against Radiation, Section 1302.

furthermore employing the guidance promulgated by the BEIR VII committee for a whole-body exposure, it follows that one fatal cancer would be
expected to result from the exposure if 3600 people inhaled radioactive
Kr-85 in the manner described. In sum, it appears that the regulation may
fail – by a factor of hundreds – to meet the protective standard it intends. The
LNT model asserted by the authorities is not representative of the physical
reality of low-level exposures.

Summary Points

1. Nature judges the probability of biological injury caused by exposure to
   ionizing radiation on an ensemble-averaged basis. Common sense does
   not address the question of "how often" events occur very well.
2. There is no such thing as a low-level exposure. All exposures are com-
   posed of discrete, high dose rate events of very brief duration.
3. Our scientific understanding of protracted exposures, including background radiation, may be incomplete, or even incorrect. A hypothesis
   known as "shot noise in radiobiological systems" has been proposed.

References

Datesman, A. (2016). Shot noise in radiobiological systems. *Journal of Environmental Radioactivity, 164*, 365–368.

Datesman, A. M. (2019). Shot noise explains the Petkau 22Na+ result for rupture of a model phospholipid membrane. *Health Physics, 117*, 532–540.

Exelon Generation. (2018). *Annual radioactive effluent release report, 2018, Oyster Creek Generating Station*. In, 22. Nuclear Regulatory Commission.

Greene, G. (2003). *The woman who knew too much: Alice Stewart and the secrets of radiation*. University of Michigan Press.

Kalinowski, M. B., & Tuma, M. P. (2009). Global radioxenon emission inventory based on nuclear power reactor reports. *Journal of Environmental Radioactivity, 100*, 58–70.

National Research Council. (2006). *Health risks from exposure to low levels of ionizing radiation*: BEIR VII phase 2. The National Academies Press.

Petkau, A. (1972). Effect of 22 Na+ on a phospholipid membrane. *Health Physics, 22*, 239–244.

Rossi, H. H. (1986). Radiation quality. *Radiation Research, 107*, 1–10.

Walker, J. S. (2000). *Permissible dose: A history of radiation protection in the twentieth century*. University of California Press.

Chapter 8
New Nuclear Power: Expensive, Slow, and Inferior

The great wind-turbine on a Vermont mountain proved that men could build a practical machine which would synchronously generate electricity in large quantities by means of wind-power. It proved also that the cost of electricity so produced is close to that of the more economical conventional methods. And hence it proved that at some future time homes may be illuminated and factories may be powered by these new means. (Putnam, 1948)
— Vannevar Bush, architect of the scientific establishment in the U.S.

In the fossil fuel era, the sun has been largely ignored. No nation includes the sun in its official energy budget, even though all the other energy sources would be reduced to comparative insignificance if it were. We think we heat our homes with fossil fuels, forgetting that without the sun those homes would be –240 degrees Centigrade ... No country uses as much energy as is contained in the sunlight that strikes just its buildings. (Hayes, 1983)
— Denis Hayes, second director of the Solar Energy Research Institute.

The Initial Adoption of Nuclear Power

The topic of this chapter is the collision in the present day between competing technologies for the generation of electrical energy: nuclear, solar photovoltaic, and onshore and offshore wind energy. The conflict has not arisen suddenly. In fact, its most prominent expression probably took place in the 1970's, when the United States embarked on the third major energy transition in its history.

While wind energy possesses a centuries-long heritage (for grinding grain, pumping water, and even generating electricity at homestead scale), its history as a technology suitable for utility-scale generation of electrical

---

Aaron Datesman is the primary author of this chapter.

D. Brugge, A. Datesman, *Dirty Secrets of Nuclear Power in an Era of Climate Change*, https://doi.org/10.1007/978-3-031-59595-0_8

Fig. 8.1 (Left) The Smith-Putnam wind turbine built on Grandpa's Knob in Castleton, VT, which commenced operation in October 1941. (Source: Putnam, 1948) (Right) Denis Hayes, organizer of the first Earth Day in 1970. (Source: AP)

energy traces to a single installation, pictured in Fig. 8.1: the 1.25 MW Smith-Putnam wind turbine, constructed in the early 1940's. Built by MIT engineer Palmer C. Putnam, heir to the Putnam publishing house, it was the largest wind turbine anywhere in the world until 1979. To design and build this pioneering facility, Putnam enlisted the help of scientific luminaries including the aeronautical engineer Dr. Theodore von Kármán of Caltech, and Dr. Vannevar Bush of MIT, the visionary of American scientific dominance in the second half of the twentieth century. Wind energy began to contribute to the US electrical grid at utility scale in 1981, with the construction of wind energy facilities in the Altamont Pass, near San Jose in California.

There are many candidate milestones that could be chosen to indicate the advent of solar photovoltaic technology. In the author's opinion, the best choice is the 1954 Bell Laboratories patent of the silicon solar cell.[1] The first utility-scale (meaning larger than 1 MW) solar photovoltaic plant was constructed by the Atlantic Richfield Oil Company in San Luis Obispo County, CA, in 1983.

---

[1] The patent, awarded in 1957, has the unremarkable title "Solar Energy Converting Apparatus".

For nuclear power technology, one might choose as an appropriate milestone the date on which the Submarine Thermal Reactor (STR) reached criticality, in March 1953. The STR was the prototype for the powerplant of the Nautilus submarine, the first vessel to complete a submerged transit of the North Pole. A more relevant choice might instead be the first commercial generation of electrical power using nuclear technology, which commenced at the Shippingport Atomic Power Station near Pittsburgh, PA, in December 1957. The power output of this facility, which operated into the 1980's, was 60 MW.

The commercial buildout of large nuclear power stations in the United States was underway by the mid-1960's. Historians employed by the Nuclear Regulatory Commission describe the situation existing at that time as a "bandwagon market":

> The bandwagon market for nuclear power reached its peak during 1966 and 1967, exceeding, in the words of one General Electric official, "even the most optimistic estimates..." [In 1967], nuclear vendors sold 31 units that represented 49 percent of the capacity ordered. (Walker & Wellock, 2010).

The economics underpinning the frenzy, however, were adverse. The bandwagon market lost hundreds of millions of dollars for Westinghouse and General Electric, the two firms competing at that time to offer "turnkey" nuclear power stations of unprecedented scale to skeptical utilities. The effort to crack open a market that did not previously exist, however, was successful: by 1980, there were 71 operating nuclear power stations in the U.S. Increasing concern regarding the particulate air pollution released by coal-fired power plants had provided a strong motivation for utility company executives to consider switching to nuclear power technology.

The onset of commercial nuclear power in the U.S. about coincided with two other events, in some senses distinct, but also all interrelated: the 1973 crisis created by the OPEC oil embargo, as well as the advent of the modern environmental movement. The first Earth Day celebration, on April 22, 1970, was a watershed moment for the latter: an estimated 20 million Americans took part. While Senator Gaylord Nelson of Wisconsin is recognized as the founder of Earth Day, the individual most responsible for making the senator's vision a reality was a young man from a working-class background named Denis Hayes, whose photo appears in Fig. 8.1. A New York Times interview from 2020 describes the path that brought Hayes to his fateful meeting with Senator Nelson in Washington, DC, in 1969:

> [Hayes] traveled across Asia and much of Africa, Eastern Europe and the Middle East, working when he needed money for the next leg and living on peanut butter and oatmeal,

and the occasional cup of coffee loaded with all of the sugar and cream on the table ... On a
meditative night in the desert, in a state of mind heightened by his "terrible diet" and the
desert chill, "It just came together in my mind that we're animals and we didn't abide by the
principles that govern the natural world," he said.

He woke up the next morning with a purpose. "I wanted to devote my life to advancing
principles of ecology as they apply to human beings and to human communities, to human
processes." (Schwartz, 2020)

Environmentalist sentiment and activism racked up a remarkable string of
successes during the Nixon administration in the early 1970's, including the
Clean Air Act, the Clean Water Act, and the establishment of the
Environmental Protection Agency. Especially because contemporary events
aligned with environmental concerns, the promotion of alternative energy
technologies was a component of the ecological fervor of the era. For
instance, the Solar Energy Research Institute (SERI) was established by legislation in 1974 (during the Ford administration). SERI opened its doors in
Golden, CO, in 1977. The second director of SERI, appointed by President
Carter in 1979, was Denis Hayes. Today SERI is known as the National
Renewable Energy Laboratory, NREL. Its 2020 budget was $545 million.

In the 1970's, wind and solar energy technologies were not ready for
large-scale deployment. Due to $16 billion spent by the Federal government
between 1951 and 1971 to develop light water reactor technology, however,
at that time nuclear power did exist as a viable option. The lesson that nuclear
power, while viable, was not economical, has perhaps been forgotten. When
delivery of oil from the North Slope of Alaska (the Trans-Alaska Pipeline
opened in 1977) commenced, the energy crises of the 1970s began to abate,
and the imperative for nuclear power in the U.S. collapsed.

The thesis of this chapter is that, while history may not repeat itself
exactly, it does often rhyme. There are lessons from the events of four or five
decades ago that resonate today. Nuclear power competes, on a spectrum of
distinct criteria, with other sources of energy. Therefore, with the appropriate context in mind, this chapter compares the performance of competing
technologies – new nuclear power, solar photovoltaic, onshore wind, and
offshore wind – on basic criteria including cost and time to deployment, in
the present day. What is the best path forward?

The Domestic Nuclear Renaissance

The author was employed in a research laboratory supporting the Nuclear
Navy in 2005, when there was much excitement about a "Nuclear
Renaissance" that appeared to be right around the corner. At that time, the

Bush administration – firmly rooted as its key figures were in the fossil energy sector – nevertheless promoted nuclear power as a zero-carbon means of energy production, compatible with the demands of environmentalists to address climate concerns. The excitement was stimulated by significant events: for instance, the Nuclear Regulatory Commission awarded the final design certification to the new Westinghouse AP1000 pressurized water reactor (PWR) in December 2005.

Nearly 18 years later, it's apparent that the promised renaissance failed to materialize – at least, not in the United States. Only one nuclear power plant has entered operation in the U.S. since then, Watts Bar Unit 2 in Tennessee. The plant went on-line in June 2016. Prior to that date, the newest nuclear power station in the U.S. was Watts Bar Unit 1, which went on-line in May 1996. (Construction began at Watts Bar Unit 2 in 1973.) In total, the material outcome of the domestic nuclear renaissance consists of two construction sites in the southeastern U.S., where the construction of four AP1000 PWRs began in the 2009 timeframe. V.C. Summer Units 2 and 3 in South Carolina were cancelled in July 2017, after an expenditure of nearly $11 billion. Several executives responsible for the project were prosecuted for fraud; at least one served a sentence in the federal penitentiary.

In Georgia, construction on Plant Vogtle Units 3 and 4 remained ongoing fourteen years after construction began. The U.S. government has provided $12 billion in loan guarantees supporting the effort. As much as $35 billion has been spent on the construction of these facilities through mid-2023. Initial criticality was achieved in Unit 3 in March 2023, with grid connection occurring the following month. It had been expected that the reactors would enter service in 2016, at a cost of $14 billion – less than half of what was eventually spent. Plagued by delays and massive cost overruns, Plant Vogtle has at least demonstrated that it is possible to build 2000 Megawatts of electrical generation in the United States using Generation IV nuclear power plant technology. No further claim regarding the potential of next-generation nuclear technology in the U.S. has been validated by experience.

The nameplate capacity, in Megawatts (MW), measures the maximum power an energy generation facility produces at one instant in time under ideal conditions. The capacity factor describes how much energy the facility delivers in operation over an extended duration. Most nuclear power stations have a capacity around 1000 MW, equal to 1 GW (Gigawatt). Although we call the utility the "power company", the charge per kW-hr on a utility bill reflects the price of energy rather than power. An average U.S. household consumes about 10,600 kW-h (11 MW-h) of energy annually.

The Vanishing Nuclear Renaissance in Historical Context

Because offshore wind (which has not been demonstrated at a large scale domestically) will be offered as a further alternative, it is legitimate to ask whether the Plant Vogtle costs result from so-called "first of a kind" (FOAK) issues, or other concerns that have been addressed and should not reoccur. A report from MIT assesses the construction "should cost" of the next AP1000 reactor to be significantly lower than the Plant Vogtle expenditure, declining still further for the tenth unit.

The prediction of lower capital costs with increasing deployment of nuclear power is not, however, supported by historical experience. According to data from the U.S. Energy Information Administration (EIA), for 75 nuclear power plants that began construction in the U.S. from 1966 until 1977, the construction costs increased from $0.623 million/MW (in 1982 dollars) in 1966–1967 to $2.132 million/MW in 1976–1977. The latter value is equivalent to $6.7 million/MW in 2023 dollars. Because there are no plans to build more AP1000 reactors, there is no prospect of attempting to validate the prediction of a reduction in the "n-th of a kind" (NOAK) capital cost against the wisdom provided by previous experience of costs that increase over time.

The conclusion that "the market has spoken" against nuclear power is to some extent surprising within the culture, because there simultaneously exists a well-funded, vocal, significant, and often rather successful public relations campaign in favor of the technology.[2] Why does "The Market" – supposedly clear-eyed, dispassionate, and not given to the unfounded concerns regarding safety voiced by mothers and activists – not respond positively to the technophilic impulse? It is the author's view that the situation arose in part because the 1979 accident at the Three Mile Island (TMI) nuclear power station in Pennsylvania has been misinterpreted to the benefit of the industry.

The dominant view is that TMI killed the domestic nuclear power industry, but this is not in fact correct. Spiraling costs halted the buildout of domestic nuclear power at the same moment that new supplies of oil from Alaska and the North Sea resolved the oil crises of the 1970's. TMI simply gave the industry cover for its economic failure. A February 1985 article in Forbes magazine – as reliable an indicator of what the market may believe as it is possible to find – makes the case:

---

[2] The author contends there exists a "nukebro" culture analogous to Silicon Valley "techbros" – they are often the same individuals! – although in recent years the industry has made a savvy effort to promote the voices of female social media influencers, engineers, and executives.

> The failure of the U.S. nuclear power program ranks as the largest managerial disaster in business history, a disaster on a monumental scale. The utility industry has already invested $125 billion in nuclear power … only the blind, or the biased, can now think that most of the money has been well spent. (Cook, 1985).

While 67 planned nuclear power facilities were canceled from 1979 through 1988, many nuclear power plants that had begun the licensing process in the 1970's continued to come online through the early 1990's. While the NRC did tighten regulation and oversight in the wake of TMI, it continued to support additional capacity coming on-line. The cancelations were therefore principally motivated by adverse economics. The distinction is important and relevant, even today, because the conditions that made nuclear power a managerial disaster in the 1970's have not abated. The technology remains dangerous, and therefore expensive. The TMI accident camouflaged a lesson about the economics of nuclear power that, thanks to the nuclear renaissance, it has been necessary to re-learn at great expense in South Carolina and Georgia.

Construction Cost and Duration

It is desirable first to compare the competing technologies on the basis of construction cost and duration. The comparison requires that one examine facilities of similar scale, which is a complicating factor because their large power output in some ways is a positive aspect of nuclear power stations. The largest photovoltaic power plant in the U.S. as of 2023 is the Solar Star (formerly, Antelope Valley) facility, which occupies 3200 acres on the edge of the Mojave Desert near Rosamond, CA. Its power output is 579 MW, with a capacity factor of 32.8%. Its construction, which began in the first quarter of 2013, required less than 2.5 years. The cost to build the Solar Star PV facility is not perfectly transparent, due to ownership changes and the method of financing, but an estimate of $2.7 billion is reasonable.

The largest onshore wind power facility is the Alta Wind Energy Center (AWEC), located in the foothills of the Tehachapi Mountains in California, not far from the Solar Star PV facility. The AWEC facility consists of 600 turbines with a combined capacity of 1548 MW. AWEC operates with a capacity factor of 23.5%. It was constructed in 11 stages, from 2010–2014. The individual stages, each approximately 100 MW in size, required less than 1 year to build. The overall cost of construction was in the range of $2.9 billion.

As of 2023, in the United States there exist only two operating offshore wind farms, both of which are small (12 MW and 30 MW). Because the

U.S. has lagged the rest of the world in offshore wind power for many years, a domestic as-built comparison between offshore wind and the other technologies should not be made. Nevertheless, the technology should not be excluded from consideration. Consider therefore the Hornsea offshore wind farm, the world's largest, seventy-five miles off the coast of the United Kingdom in the North Sea. Two of four proposed stages have been built and are operational. Hornsea 1, with a capacity of 1200 MW, was built between 2016 and 2020 at a cost of approximately $5 billion. Its capacity factor stands at about 47%.

The comparison between nuclear, solar photovoltaic, and onshore and offshore wind energy generation is summarized in Table 8.1, which at a high level provides a reasonably direct comparison between the competing technologies at similar (though not identical) scale as they have been built out in the real world. The annual generation in MW_{avg} is given by the nameplate capacity multiplied by the capacity factor, which accounts for such considerations as the intermittency of renewable resources, curtailment, and downtime for repair and maintenance. The capacity factor for Plant Vogtle is assumed to be 90%, nearly matching that of the U.S. nuclear industry overall. (It may, one should note, require many years of operating experience to reach this level of reliability.)

The normalized costs in Table 8.1 represent the overall cost of construction divided by the energy generated. According to this simple comparison, one concludes that Plant Vogtle is somewhat more expensive than utility-scale photovoltaic technology was 8 years ago, twice as expensive as onshore wind about a decade ago, and around 75% more expensive than offshore wind installed in the North Sea a few years ago.

Table 8.1 Comparison of normalized construction costs (final column) between nuclear and renewable technologies

	Technology	Completed	Cost (bn $)	Capacity (MW)	Avg. Gen. (MW_{avg})	Norm. Cost ($mm/ MW_{avg})
Plant Vogtle Units 3 & 4	Gen IV Nuclear	2023	35	2500	2234	15.7
Solar Star	Solar PV	2015	2.7	579	190	14.2
AWEC	Onshore Wind	2014	2.9	1548	364	8
Hornsea 1	Offshore Wind	2020	5	1200	564	9

The distinction between the capacity (in MW) and the average generation (in MW_{avg}) is the capacity factor

The comparison is incomplete because it fails to account for ongoing and future costs that include fuel, operations and maintenance (O&M), decommissioning, and safe storage of spent nuclear fuel. However, since renewables have no fuel cost (and leave no lethally dangerous waste behind) the omission further privileges renewables over nuclear, only reinforcing the basic conclusion. In short, based upon what has been demonstrated in the real world it appears that nuclear is more expensive than renewable energy technologies, and as much as a decade slower to bring on-line.

Levelized Cost of Energy

Because it is not based solely upon the installation expense, the parameter known as the Levelized Cost of Energy (LCOE) provides a more complete and useful assessment of competing technologies. The LCOE addresses a bottom-line question. Since the consumer of electrical energy is mostly concerned about the cost per kW-hr appearing on a monthly invoice from the utility company, the LCOE is defined as the lifetime cost of operating a power plant divided by the energy it produces. Calculation of the LCOE yields an expense per MW-hr of energy consumed.

A 2022 analysis by researchers at Lawrence Berkeley National Laboratory (LBNL) found an LCOE for utility-scale PV of $33 per MW-hr, nearly the same as the $34 per MW-hr for onshore wind energy found by NREL researchers in 2021. The same report found LCOE values of $78 and $133 per MW-hr for fixed-bottom and floating offshore wind, respectively. Meanwhile, a 2018 report from the Energy Information Administration (EIA) found the LCOE for new nuclear power coming on-line in 2021 to be $90 per MW-hr. The information is summarized in Table 8.2.

The LCOE values summarized in Table 8.2 are mostly consistent with the normalized costs of construction given in Table 8.1, perhaps except for the

Table 8.2 Comparison of Levelized Cost of Energy (LCOE) between new nuclear power, and renewable technologies

Resource	LCOE ($/MW-hr)	Source
New nuclear power	88	EIA (U.S.EIA, 2022)
Natural gas	40	*Ibid*
Utility-scale PV	33	LBNL (Bolinger et al., 2022)
Onshore wind	34	NREL (Stehly & Duffy, 2012)
Offshore wind, fixed-bottom	78	*ibid*
Offshore wind, floating	133	*ibid*

LCOE of offshore wind. Construction is, of course, a significant component of overall lifecycle cost. The discrepancy between construction costs vs. LCOE for onshore and offshore wind is most likely attributable to higher O&M costs for offshore facilities.

The acronyms might be obscure to the reader, but it is worthwhile to point out that the organizations represented in Table 8.2 (EIA, LBNL, and NREL) are all components of the U.S. Department of Energy.[3] Therefore, the most important takeaway from the information presented in Table 8.2 ought to be that the principal advocate for the nuclear renaissance (DOE) itself acknowledges that energy generated by new nuclear power would be nearly three times more expensive than energy generated by renewable alternatives.

The Immediate Opportunity Cost of Plant Vogtle

According to statistics compiled by Bloomberg, total domestic spending on renewable energy technologies (principally onshore wind and photovoltaic, but also including spending on electric vehicles and batteries) increased from $10 billion in 2004 to $47 billion in 2007, was reasonably consistent within the range from $45 – $65 billion from 2010–2019, and increased again in 2020. In 2021, domestic spending on renewable energy technologies exceeded $100 billion annually. The cost of Plant Vogtle can be roughly, but fairly, evaluated within the context of this information. Though spread out over 14 years, the construction of the nuclear power facility in Georgia consumed 6–9 months of total domestic investment in the renewables sector. This allocation of resources, supported by substantial loan guarantees from the U.S. government, purchased nothing but delay. Plant Vogtle Unit 3 did not deliver a single kW-hr of energy to the grid until 2023.

What if the money spent to construct Plant Vogtle Units 3 and 4 had instead been invested in renewable energy facilities? This is a reasonable means by which to assess the opportunity cost associated with Plant Vogtle. The annualized cost of construction of the two AP 1000 reactors at the site lies in the range of $2.5 billion, an amount coincidentally about equal to the total cost of the individual photovoltaic and wind energy facilities listed in Table 8.1.

If the money spent to build Plant Vogtle had been, instead, allocated toward building similar photovoltaic and wind energy facilities beginning in 2014 or 2015, by 2023 it would have been possible to deploy six completed

---

[3] One might make the distinction that the latter two are independent entities operated on behalf of DOE.

facilities like either Solar Star or AWEC. In the first case the deployed average generation would have been 1140 MW_{avg} at a cost of \$16 billion; in the second, 2184 MW_{avg} at a cost of \$17 billion.[4] The renewable energy facilities, unlike Plant Vogtle, would have begun to push power on to the electrical grid as early as 2015, with more coming on-line every subsequent year. Plant Vogtle Units 3 and 4 together cost \$35 billion, pushed no energy on to the grid until 2023 (at which point Unit 4 had still not come on-line), and will (optimistically) provide average generation of 2234 MW_{avg}.

On the basis of rapid deployment and cost – both installation cost, and LCOE – it is clear that utility-scale photovoltaic and onshore wind today each remarkably outperform new nuclear installations at current levels of penetration. It is very difficult to make predictions, however, especially about the future. To consider one example, the comparison between renewables and new nuclear power is not necessarily stationary as additional resources come on-line. The best sites for renewable energy generation have naturally been built out first, so that capacity factors of new facilities will decrease (and LCOE rise) as penetration increases. How will the situation appear once the grid has been built out to 50% powered by wind and solar energy? The question requires thoughtful analysis beyond the scope of this discussion.

The Current Situation

A facility costing billions of dollars to construct cannot simply be copied. Among other considerations, siting is an important issue for deployment. Although copies cannot be made, facilities like AWEC and Solar Star are being built, and at an astonishing rate. Figure 8.2 illustrates capacity additions for wind and solar energy installations in the U.S. on an annual basis from 2010–2021.

Wind energy installations contributed new nameplate capacity in the range from 7–9 GW every year from 2015–2019, while solar energy nameplate capacity grew by more than 10 GW every year from 2016–2020. New wind capacity in 2021 was 13 GW, while in 2021 new solar photovoltaic capacity was 23.6 GW. If the average capacity factor of these photovoltaic installations is 20%, then in just one recent year solar energy added electrical energy generation about equivalent to <u>five</u> AP1000 nuclear reactors to the grid.

---

[4]The estimates are pessimistic because, using the information from Table 8.1, 2014–15 costs are assumed. For solar photovoltaic particularly the costs have declined steadily and substantially.

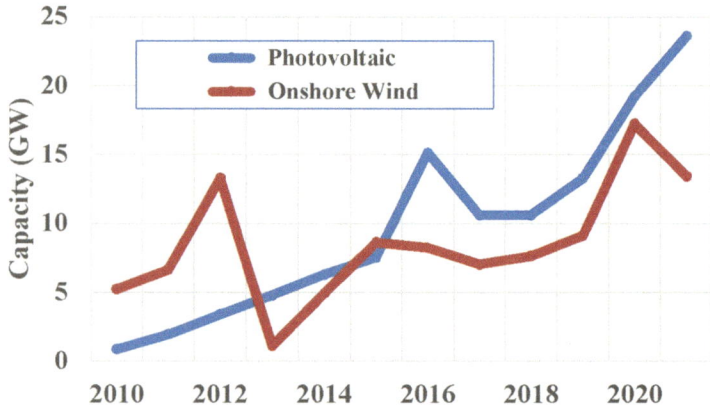

Fig. 8.2 Historical additions to generating capacity by renewable sources. The data are from Bloomberg. (Nathalie, 2023)

In 2023, EIA expects the grid to add capacity equal to 29.1 GW from utility-scale solar photovoltaic, 9.4 GW from battery storage, 7.5 GW from natural gas, 6 GW from wind power (including 130 MW of offshore wind), and 2.2 GW from nuclear power (as the Plant Vogtle reactors come on-line). Overall, utility-scale solar photovoltaic represents more than half (54%) of forecasted additional capacity. The situation for renewables today, in short, echoes the favorable situation that existed for nuclear power in 1967, when nuclear technology accounted for 49% of new capacity additions.

Today, however, no additional capacity from nuclear power is likely to be constructed after the completion of Plant Vogtle Units 3 and 4. The market has given a very clear signal that renewable technologies are a good and promising investment, while nuclear power is disfavored.

The Future is Offshore[5]

While in 2023 there were only two small offshore wind farms operating domestically, the U.S. possesses a tremendous potential for developing offshore wind. The most recent update available from NREL finds an offshore resource of 1476 GW that could be accessed using fixed-bottom turbines (possible only in relatively shallow water), and 2773 GW using floating turbines. By way of comparison, according to EIA the total utility-scale

---

[5] It is appropriate to disclose that the author was employed as a contractor supporting the Offshore Wind Initiative in the wind energy program within DOE for about a year, circa 2012.

capacity for the generation of electrical energy in the United States at the end of 2022 was 1160 GW.[6] That is, the sum of all of the electrical power consumed within the United States is less than the estimated resource from offshore wind.

Worldwide (mostly in the United Kingdom, China, and Germany), the capacity of installed offshore wind power exceeded 35 GW in 2020. In the U.S., it seems conceivable that a remarkable transformation could be on the horizon. As of August 2022, there was a construction pipeline for offshore wind projects of 40 GW; twenty-four power purchase agreements for off-shore wind power, totaling 17.6 GW, were already in effect.

Deployment may accelerate from 2023–2030, in response to the goal of the Biden administration to install 30 GW of offshore wind power by 2030. Whether this goal is met, of course, remains to be demonstrated. The grow-ing success of the onshore wind industry in the U.S., the vast scale of the resource, and the experience that exists with offshore wind in Europe and China, taken together all suggest that the main barriers are only political and social.

The path forward toward 80% and then 100% renewable energy will therefore significantly include offshore wind, which is already clear due to the scale of existing power purchase agreements for wind energy facilities that have yet to be built. However, the technology lacks a strong domestic track record, and is not obviously superior to new nuclear power on level-ized cost of energy. Why would it make sense to pursue one approach over the other? That is: while financially it does not make sense to build new nuclear power plants today, we might wish that we had when renewable penetration has reached 50%.

In the author's opinion, this is unlikely to be the case. Nuclear power plants are expensive because the technology is inherently dangerous, few are built, and the necessary components must be engineered to an astound-ing level of quality. History demonstrates that predicted cost savings do not materialize. Photovoltaic panels, on the other hand, have obeyed the rule of mass production. The cost of a solar panel declined by a factor of eight in about a decade: from $2.15 per Watt in 2010, to $0.27 per Watt in 2021. While wind energy facilities will not show the same dramatic cost declines exhibited by photovoltaic components, the rules of mass production will also apply. The LCOE of offshore wind will likely decline, making new

---

[6] Small-scale photovoltaic installations contributed an additional 39.4 GW in 2022. Assuming a capacity factor of 20%, the energy generated by this means is equivalent to nearly eight nuclear power stations. As of 2023, there are 93 operating nuclear power stations in the U.S.

nuclear power facilities economically uncompetitive with this vast, and untapped, resource.

The U.S. Department of Energy forecasts that, by 2035, LCOE will decline to $53 per MW-hr for fixed-bottom, and $64 per MW-hr for floating, offshore wind energy. The reductions will occur due to efficiencies which arise as the global deployment of offshore wind energy increases by a factor of nine, over a span of only a dozen years. The forecasted horizon to realize these cost savings is shorter than the construction duration for Plant Vogtle after breaking ground in 2009. New nuclear power facilities, in short, are the wrong response to climate change: too slow, too expensive, and a misallocation of financial resources away from superior alternatives.

The Long-Term Opportunity Cost

The discussion in this chapter opened with the historical view, beginning with the construction of a utility-scale wind energy facility on a mountain-top in Vermont in the early 1940's. Nuclear power, wind energy, and solar photovoltaic technologies all already existed in nascent form in the years circa 1940–1955, decades before "renewable energy" or "sustainability" appeared in discourse. The development of nuclear power to create a commercial technology from modest beginnings benefited both from institutional inertia[7] due to the massive wartime effort to produce the atomic bomb, as well as national security concerns at the outset of the Cold War. Research and development (R&D) of the competing renewable technologies, on the other hand, was not meaningfully supported until the establishment of the Department of Energy in 1978. The funding landscape in the years from 1948–2018 is summarized in Fig. 8.3.

It is clear from Fig. 8.3 that funding for nuclear and fossil energy has greatly exceeded funding for renewables (and energy efficiency) over a time scale of decades. In recent years, the data indicate that there has been approximate parity. From fiscal year FY09 − FY18, the Congressional Research Service reports expenditures of $9.4 billion for Renewables, $8.2 billion for Energy Efficiency, $10 billion for Fossil Energy, and $13.7 billion for Nuclear Energy. The Federal R&D expenditure for nuclear energy over a seventy-year horizon has been $110 billion.

The Federal appropriation for SERI in 1980 was $130 million, estimated to exceed the combined amounts spent on renewables research and

---

[7] The author likes to remark that the Manhattan Project never died, it just became the Department of Energy.

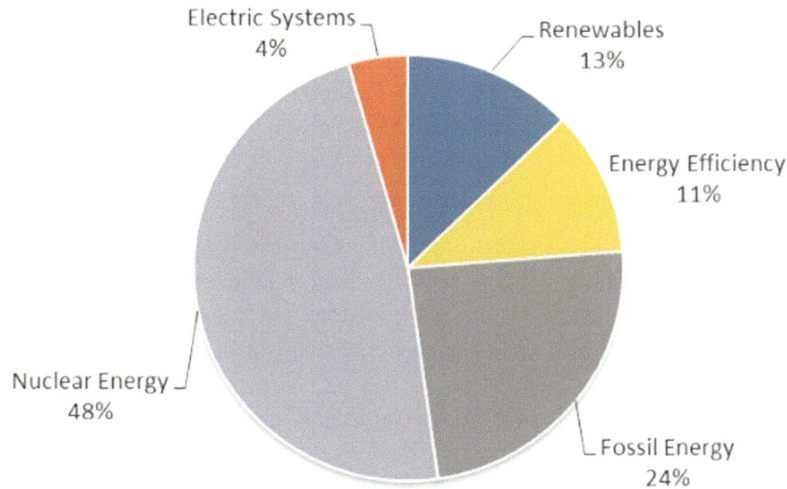

Fig. 8.3 Federal government support for energy technology research from 1948–2018, in 2016 dollars. Spending in the Renewables category prior to the establishment of the Department of Energy in 1978 was rather small, less than $2 billion. This information was compiled by the Congressional Research Service. (Clark, 2018)

development by every other country in the world at that time. President Carter lost the election of 1980, however, and in 1981 the Reagan administration reduced funding for the new facility to only $30 million. About half of the staff were released. Denis Hayes resigned his position as director of SERI in protest. Hayes reflected on the situation in an interview conducted in 2012, a point in time when solar photovoltaic capacity was being added at only about one-tenth of the current rate:

> As a planet, we're now moving the way we would have if we'd gotten Carter re-elected. The tragedy is that the United States dropped the torch. The flame was kept alive by Japan and then Germany with their feed-in tariffs. A dozen other countries now have worked to achieve the volume we needed to drive us down the cost curves. The basic technology of what we do now was all within our grasp in the Carter years. (Masia, 2012)

The $110 billion in Federal R&D funding for nuclear energy, like the money spent to construct two AP1000 reactors at Plant Vogtle, should also be considered a statement of opportunity cost. Recognizing that the competitive landscape between new nuclear power and renewables strongly favors renewables today, where might we be if the decades of support lavished upon a "managerial disaster" had instead been allocated to patient, long-term development of wind energy and solar photovoltaics? The society cannot go back in time, of course, but wiser decisions can (and must) be made moving forward. The quickest way to reduce the quantity of carbon released into the atmosphere is to deploy wind and solar photovoltaic energy generation facilities at vast scale and as quickly as possible. Nuclear energy is an impediment to this urgent goal.

Summary Points

1. New nuclear power is more expensive than renewable alternatives, and far slower to build and bring on-line. Renewable technologies (onshore and offshore wind, and solar photovoltaic) are the best means to reduce the amount of carbon being released to the atmosphere as quickly as possible.
2. Nuclear power has never been economical in a domestic context. On a generational basis, its promoters have been able to advance deployment of the technology in response to societal crises.
3. The Federal government has supported research and development of nuclear power technology on a vast scale over decades. At the present time, R&D for renewables remains less than that supporting nuclear power technology.

References

Bolinger, M., Seel, J., & Warner, C., Robson, D. (2022). *Utility-scale solar, 2022 edition: Empirical trends in deployment, technology, cost, performance, PPA pricing, and value in the United States.* Lawrence Berkeley National Laboratory. https://emp.lbl.gov/publications/utility-scale-solar-2022-edition

Clark, C. E. (2018). *Renewable energy R&D funding history: A comparison with funding for nuclear energy, fossil energy, energy efficiency, and electric systems R&D.* Congressional Research Service (CRS). https://crsreports.congress.gov/product/pdf/RS/RS22858

Cook, J. (1985). Nuclear follies. *Forbes, 135*(3), 82.

Hayes, D. L. (1983). Environmental benefits of a solar world. In D. Rich, J. M. Veigel, A. M. Barnett, & J. Byrne (Eds.), *The solar energy transition: Implementation and policy implications* (p. 186). Avalon Publishing.

Masia, S. (2012). *Hayes: It's within our grasp to transition quickly.* American Solar Energy Society. https://ases.org/hayes-its-within-our-grasp-to-transition-quickly/

Nathalie, L. (2023). *Sustainable energy in America 2023 factbook. Tracking market and policy trends.* Bloomberg Finance L.P.

Putnam, P. (1948). *Power from the wind.* Van Nostrand Reinhold.

Schwartz, J. (2020). *The 'Profoundly Radical' message of earth day's first organizer.* New York Times Company.

Stehly, T., & Duffy, P. (2012). *2021 Cost of wind energy review.* National Renewable Energy Laboratory. https://www.nrel.gov/docs/fy23osti/84774.pdf

U.S. EIA. (2022). *Levelized costs of new generation resources in the annual energy outlook 2022.* U.S. Energy Information Administration. https://www.eia.gov/outlooks/aeo/pdf/electricity_generation.pdf

Walker, S. J., & Wellock, T. R. (2010). *A short history of nuclear regulation* (p. 27). U.S. Nuclear Regulatory Commission. https://www.nrc.gov/docs/ML1029/ML102980443.pdf

Correction to: Climate Change: Melting Ice and Statistical Models

Correction to:
Chapter 1 in: D. Brugge, A. Datesman,
Dirty Secrets of Nuclear Power in an Era of Climate Change,
https://doi.org/10.1007/978-3-031-59595-0_1

The original version of the chapter "Climate Change: Melting Ice and Statistical Models" was inadvertently published with incorrect image panel arrangement in Figure 1.1. The correction has been updated in the chapter.

The updated version of this chapter can be found at
https://doi.org/10.1007/978-3-031-59595-0_1

D. Brugge, A. Datesman, *Dirty Secrets of Nuclear Power in an Era of Climate Change*, https://doi.org/10.1007/978-3-031-59595-0_9

Afterword

> Canadian wildfires have this year burned a land area larger than 104 of the world's 195 countries. The carbon dioxide released by them so far is estimated to be nearly 1.5 billion tons – more than twice as much as Canada releases through transportation, electricity generation, heavy industry, construction and agriculture combined. In fact, it is more than the total emissions of more than 100 of the world's countries – also combined.
>
> – David Wallace-Wells
>
> https://www.nytimes.com/2023/09/06/opinion/columnists/forest-fires-climate-change.html

In the year in which we were writing this book, 2023, wildfires erupted across Canada at a scale not seen before. Smoke from the fires traveled hundreds of miles, blotting out the sun in cities across the northern United States and threatening the health of millions of vulnerable people. It was a sobering reminder that the feedback loops that might be initiated by climate change have not been entirely foretold. It was also yet another indication of the need to act quickly to avoid the worst impacts of climate change.

It is critical that readers of this book who are concerned about climate change understand that we share their concern. If anything, we might err on the side of greater concern than establishment, mainstream science on this issue. Thus, when we critique nuclear power and come to the conclusion that it is not a viable or responsible response to the climate change problem, we do so fully aware of the seriousness of the situation.

We support a nuanced, complex, and evidence-based assessment of problems and solutions. It is in that spirit that we wrote this book, trying hard not to put our finger on the scale to tilt the evidence toward a preconceived and desired outcome, but rather to assess it fairly. For example, we do not support a rapid phase out of nuclear, especially if it is replaced by burning fossil fuels. That said, we, like everyone, are human and have our biases based on experience and what we have been taught and exposed to. Nevertheless, as best we could, we sought to be fair to nuclear power relative to the dangers of climate change.

D. Brugge, A. Datesman, *Dirty Secrets of Nuclear Power in an Era of Climate Change*, https://doi.org/10.1007/978-3-031-59595-0

Thus, it is notable that it appears that the proverbial nail in the coffin of a nuclear renaissance, in high income countries anyway, is not science and risk per se, but rather the verdict of the market. We are both scientists and skeptical of unfettered markets driving decisions, yet here we are, acknowledging that at the end of the day, capitalist economic imperatives render nuclear unviable.

Although it is not the focus of this book, it is notable that there are also downsides to solar and wind, most obviously the devastating local impacts from mining rare earth metals. But we are not purists. While we feel those impacts must be reduced and the affected populations compensated for any harms, unlike nuclear, we think the net benefit and potential of wind and solar are positive. It is rare in this world that there are no tradeoffs.

In closing, we implore the reader to think for themselves, to join in efforts to avert the impacts of climate change, and to be willing, where appropriate, to support the resources necessary to mitigate the ancillary harms of what we need to do to avoid larger damage.

Index